50 Years of Materials Science in Singapore

World Scientific Series on Singapore's 50 Years of Nation-Building

Published

50 Years of Social Issues in Singapore
    edited by David Chan

Our Lives to Live: Putting a Woman's Face to Change in Singapore
    edited by Kanwaljit Soin and Margaret Thomas

50 Years of Singapore–Europe Relations: Celebrating Singapore's Connections with Europe
    edited by Yeo Lay Hwee and Barnard Turner

Perspectives on the Security of Singapore: The First 50 Years
    edited by Barry Desker and Cheng Guan Ang

50 Years of Singapore and the United Nations
    edited by Tommy Koh, Li Lin Chang and Joanna Koh

50 Years of Environment: Singapore's Journey Towards Environmental Sustainability
    edited by Tan Yong Soon

Food, Foodways and Foodscapes: Culture, Community and Consumption in Post-Colonial Singapore
    edited by Lily Kong and Vineeta Sinha

50 Years of the Chinese Community in Singapore
    edited by Pang Cheng Lian

Singapore's Health Care System: What 50 Years Have Achieved
    edited by Chien Earn Lee and K. Satku

Singapore–China Relations: 50 Years
    edited by Zheng Yongnian and Lye Liang Fook

Singapore's Economic Development: Retrospection and Reflections
    edited by Linda Y. C. Lim

50 Years of Urban Planning in Singapore
    edited by Chye Kiang Heng

Singapore's Real Estate: 50 Years of Transformation
    edited by Ngee Huat Seek, Tien Foo Sing and Shi Ming Yu

The Singapore Research Story
    edited by Hang Chang Chieh, Low Teck Seng and Raj Thampuran

The complete list of titles in the series can be found at
http://www.worldscientific.com/series/wss50ynb

World Scientific Series on
Singapore's 50 Years of Nation-Building

50 YEARS OF MATERIALS SCIENCE IN SINGAPORE

Editors

Freddy Boey
B. V. R. Chowdari
Subbu S. Venkatraman

Nanyang Technological University, Singapore

World Scientific

NEW JERSEY · LONDON · SINGAPORE · BEIJING · SHANGHAI · HONG KONG · TAIPEI · CHENNAI · TOKYO

Published by

World Scientific Publishing Co. Pte. Ltd.
5 Toh Tuck Link, Singapore 596224
USA office: 27 Warren Street, Suite 401-402, Hackensack, NJ 07601
UK office: 57 Shelton Street, Covent Garden, London WC2H 9HE

Library of Congress Cataloging-in-Publication Data
Names: Boey, Freddy, editor. | Chowdari, B. V. R., editor. | Venkatraman, Subbu, editor.
Title: 50 years of materials science in Singapore / editors, Freddy Boey (Nanyang Technological University,
  Singapore), B.V.R. Chowdari (National University of Singapore), Subbu Venkatraman
  (Nanyang Technological University, Singapore).
Other titles: Fifty years of materials science in Singapore | World Scientific series on
  Singapore's 50 years of nation-building.
Description: Singapore ; Hackensack, NJ : World Scientific Publishing Co. Pte. Ltd., [2016] | 2016 |
  Series: World Scientific series on Singapore's 50 years of nation-building |
  Includes bibliographical references and index.
Identifiers: LCCN 2016000585| ISBN 9789814730693 (hardcover ; alk. paper) |
  ISBN 9814730696 (hardcover ; alk. paper)
Subjects: LCSH: Materials science--Research--Singapore--History.
Classification: LCC TA402.512.S55 A15 2016 | DDC 620.1/1095957--dc23
LC record available at http://lccn.loc.gov/2016000585

British Library Cataloguing-in-Publication Data
A catalogue record for this book is available from the British Library.

Copyright © 2016 by World Scientific Publishing Co. Pte. Ltd.

All rights reserved. This book, or parts thereof, may not be reproduced in any form or by any means, electronic or mechanical, including photocopying, recording or any information storage and retrieval system now known or to be invented, without written permission from the publisher.

For photocopying of material in this volume, please pay a copying fee through the Copyright Clearance Center, Inc., 222 Rosewood Drive, Danvers, MA 01923, USA. In this case permission to photocopy is not required from the publisher.

In-house Editor: Christopher Teo

Typeset by Stallion Press
Email: enquiries@stallionpress.com

Foreword

In "*Uncle Tungsten*" the joy of discovering "the secrets of alchemical realities", through materials science, was vividly described by the gifted author and scientist, Oliver Sacks.

Exoticism found in materials like rubidium, tungsten, tantalum and (my favourite) titanium is described in the context of the history of chemistry and the advancement of science. It is these same wonders and advancement of science, through the lens of materials research and education, which this book chronicles as a passage in the course of Singapore's economic development and societal progress.

The aspiration to be among the World's best in the field of materials science is manifest in the impressive vision and leadership of the Nanyang Technological University to build the largest School in this discipline. Over two hundred and fifty students graduate each year, contributing to various facets of our economy and society. Materials research is also a pan-Singapore effort with world class scientific pursuits at the National University of Singapore and Research Institutes from the Agency for Science, Technology and Research.

It is this remarkable breadth, in an island as small as Singapore, which has contributed to excellence in fields that span electronics to biomedical science, energy to water. These are areas where progress in materials science and their applications are important to push the boundaries of possibilities and future advances.

This book describes many examples of intrepid scientists who explore the frontiers and give us a glimpse of how a fundamental understanding of nature's orchestration of atoms — which is the essence of materials science — and the ability to manipulate and engineer these systems to solve some of the most challenging problems in technology today make them some of our modern day "heroes".

Beyond the magic of scientific discoveries are clear connections made to economic development. In the area of biomedical sciences, devices have been developed from the origins of advances in materials where global companies have been formed and lives can be improved. The reader will find that significant effort has been made in solar energy, energy storage and membrane technologies for desalination and remediation. This is profoundly important work aimed at mitigating climate change through the use of clean energy and our need for sufficiency and resilience in the vital supply of water for Singapore.

Being trained in the field, materials science to me is full of adventures and exploration of unknown territories designed by the hand of nature in the universe of matter. As Freddy Boey states "we probably only know 10%. The other 90% is still out there waiting to be invented or discovered". In this book we find discoveries and dedication, exploration and excitement, by top calibre practitioners who are at the vanguard of a premier field in science.

I thank the editors Freddy Boey, B.V.R. Chowdari and Subbu Venkatraman for their diligence and commitment to share the work of pioneers who have narrated Singapore's work in materials science — one of international standing and global importance.

Dr. Raj. Thampuran
Managing Director
Agency for Science, Technology and Research.

Contents

Foreword		v
About the Editors		ix
Overview		xi
Chapter 1	Historical Narrative Early Beginnings to Present *Freddy Boey*	1
Chapter 2	Composites, Nanocomposites and Hybrid Materials *Chaobin He, Xiao Hu, Zhang Yu and John Wang*	21
Chapter 3	Materials for Water Remediation (Membranes) *Sui Zhang, Lin Luo, Zhi Wei Thong and Tai-Shung Chung*	37
Chapter 4	Nanostructured Catalytic and Adsorbent Materials for Water Remediation *Zhong Chen and Teik Thye Lim*	75
Chapter 5	Solar Energy and Energy Storage Materials and Devices Research in Singapore *D. Sabba, J. Wang, M. Srinivasan, A.G. Aberle and S. Mhaisalkar*	113
Chapter 6	50 Years of Biomaterials Research in Singapore *Subbu Venkatraman, Swee Hin Teoh and Ali Miserez*	157
Chapter 7	2D Materials *Andrew T. S. Wee, Kian Ping Loh and Antonio H. Castro Neto*	179
Chapter 8	Electronic Materials Research in Singapore *Chee Ying Khoo, Pooi See Lee, Sze Ter Lim and Chee Lip Gan*	197
Chapter 9	"Singaporean" Materials Science: What does the Future Hold? *Subbu Venkatraman*	225
Index		229

About the Editors

Prof. Boey is a pioneer in the use of functional biomaterials for medical devices. He has developed 110 patents and founded several companies to commercialize his cardiovascular, ocular and surgical implants. Among his inventions is a surgical tissue retractor that has been sold in the US, Europe, Japan, Middle East and India. His customizable hernia mesh is the first such surgical mesh approved for sale by the US FDA and his company Amaranth Medical which develops full biodegradable heart stents, has recently received a buy in by Boston Scientific. His most recent company, Peregrine, is based on a nano based drug delivery system to treat glaucoma, which has been successfully deployed in human trials.

Apart from serving as Director on the boards of the Health Science Authority Singapore and Defense Science Organization National Laboratories, Prof. Boey also contributes on the boards of several nationally-funded research centres. He is a Fellow of the Institute of Materials in the UK, a founding Fellow of the Singapore Academy of Engineers, a Fellow of the Institute of Engineers Singapore, and a member of the National Cybersecurity R&D Laboratory (NCL).

Prof. Chowdari was awarded "Officer in the Order of Academic Palms" title by French Government, "Business Event Ambassador" title by Government of Singapore, and "Outstanding Science Entrepreneur" award by NUS. He is the elected member of the "ICSU Regional Committee for Asia and the Pacific" and also the Academician and Vice-President of the "Asia Pacific Academy of Materials". He is appointed as the Honorary Advisor to the Government of Andhra Pradesh,

India. He has also served as the President of the International Union of Materials Research Societies. Currently he is the President of the Materials Research Society of Singapore. His research interests deal in energy storage devices including lithium ion batteries.

Prof. Subbu Venkatraman has a PhD in Polymer Chemistry from Carnegie-Mellon University. He is the Chair of Materials Science and Engineering at NTU, which is currently ranked #2 worldwide by US News & World Report. He has spent about 15 years in biomedical R&D in the USA, working with various applications of polymeric biomaterials. He held a senior position in R&D at Alza Corporation prior to joining NTU as Associate Professor in 2000. He has over 60 patents in the field of biomedical materials, and his inventions have led to the co-founding of two companies, Amaranth Medical and Peregrine Ophthalmic, for cardiovascular and ocular devices. His work on nanomedicine for glaucoma has led to a joint Presidential Technology Award in 2014.

Overview

In 2011 Thomson Reuters published its first topical review devoted to Materials Science and Technology, and by doing so enhanced its traditional scope with focus on geographical reports. One of the up-front key statements of this report concluded that "The 21st century may well bring forth a new era, one of revolutionary discoveries in materials research that result in far-reaching changes for society and how we live." When one now compares the world's share of papers on materials indexed in Web of Science, it becomes evident that the output of this discipline is the highest in exact natural and engineering sciences, and has exceeded that of chemistry, conventional engineering, or physics. Striking is to observe the tremendous growth in materials research output in the Asia-Pacific region, which by now has surpassed the output of the EU-15 countries by a factor of two, and that of the US by more than a factor of three. Singapore has had its leading share in this growth, including materials research at Singapore's leading universities, like NTU and NUS, as well as in the Research Institutes of Singapore's Agency for Science, Technology and Research, A*STAR. The report entitled "50 Years of Materials Science in Singapore" published by World Scientific, authored by some of Singapore's best materials scientists, and edited by some of Singapore's materials science leaders including Professors Freddy Boey, B.V.R. Chowdari and Subbu S. Venkatraman, begins with a great overview of the history and choices made by Singapore over the last decades to bring the country to the top of science, technology and innovation. The volume truly demonstrates the tremendous growth in materials science and technologies using case studies taken from Singapore's finest contributions. These case studies cover topics like (nano)composites and hybrid materials, water purification membranes, water treatment, materials for energy storage and devices, biomaterials, 2D functional as well as electronic materials. For an aficionado in materials it is clear that most of these topics are in the focus of THE LIST of contemporary global challenges for materials science and technology. The report very successfully combines science history and science politics with research reports having a great intellectual depth, and it also makes a very pleasant read. It is a useful and highly recommended volume for scientists, science decision makers, technologists and managers alike. By outlining cutting edge materials research embedded in the context of Singapore's science and technology history it will be a very helpful read also to choose the best way to move forward with future materials.

G. Julius Vancso FRSC, MHUS
Professor of Materials Science and Technology of Polymers,
University of Twente, the Netherlands
Visiting Professor of Materials Science and Engineering,
Nanyang Technological University, Singapore

1 Historical Narrative Early Beginnings to Present

Freddy Boey*

Introduction

In her remarkable fifty years since independence, Singapore has had exceptional achievements in many areas, none so more than in Education and Research. The very first formal Materials Science programme was started in 1991 when the Division of Materials Engineering was formed under the School of Applied Science in NTU. Then in 1996, NUS started its Materials Science programme. Today, both NTU and NUS have not only developed strong MSE undergraduate and PhD programmes, they have also been both ranked top 10 in the world by the QS Subject World Ranking. US News had even ranked NTU second in the world only behind MIT, an incredible achievement in less than 25 years.

Backed by significant investments made by the government, Singapore is now a world-class centre for research and development (R&D). This has been supported by the four pillars: an educated and skilled workforce; a well-connected and integrated public sector; a supportive government, business, and regulatory environment; and government-supported research institutes that develop both basic and translational R&D capabilities. Much like other areas of study, materials science in Singapore had benefitted from a friendly research environment, an ecosystem that had proven to be conducive and resilient in furthering our understanding of materials and their properties and putting that understanding to good use.

MSE research had also since helped Singapore to create innovative technology in aerospace, microelectronics, biomedical devices, energy, water and manufacturing. The most exciting aspect of materials science is that there is no end to what we can learn, and what we can achieve. It is believed that of all the possible materials systems that can be discovered or invented in the whole world, we probably only know about 10 %. The other 90% is still out there waiting to be invented or discovered.

---

*Provost and Deputy President, Nanyang Technological University.

The Beginnings

Since attaining her independence in 1965, Singapore has surged forward in her efforts towards industrialisation and has done so successfully. At that time the two pivotal decisions that shaped Singapore's economic future were the establishment of the Jurong Industrial Estate and the Economic Development Board Ordinance of 1961.

Geographically, Singapore's main economic activities were based around the port area (Evers 1991). In 1965 manufacturing was mostly limited to low-skilled areas such as kitchenware and wigs, among others. In a short span of time, momentum and diversity in manufacturing production picked up and reached an annual average growth rate of almost 16% between 1959 and 1967. It further rose to 31.8% from 1967 to 1971, on the basis of gross value-added. By 1971, the gross value-added at current prices was almost 10 times that of 1959, the highest being in the petroleum and chemical industries, followed by the electronics and electrical machinery and apparatus, shipbuilding and repairing, and engineering industries.[1]

The dynamics of the nation's economy was impacted by a booming manufacturing sector. The industry sector that included manufacturing, construction, quarrying, electricity, gas, and water services contributed to about a third of GDP in 1971 from just 13% in 1959. By 1971, manufacturing employed more people that other sectors. This was important because these developments were changing the direction of the labour force in Singapore.[2]

Singapore's focus on areas of R&D has always been dictated by the economic needs of the country. As such, it has always been based on application, and use to the industries that were being promoted as the cornerstone of Singapore's economic development in that period of time. These immediate and future industries were

Breakdown of Gross Domestic Product by origin[3]		
	1959	1971
Industry	13.0%	30.9%
Commerce	31.7%	29.5%
Services	49.1%	36.7%
Agriculture & Fishing	6.2%	2.9%

[1] Some basic problems of industrialisation in Singapore. Author: SY Lee. Journal of Developing Areas, Vol. 7, No. 2 (Jan., 1973), pp. 185–216.
[2] Some basic problems of industrialisation in Singapore. Author: SY Lee. Journal of Developing Areas, Vol. 7, No. 2 (Jan., 1973), pp. 185–216.
[3] ibid.

identified by the government as industries where present and future employment and investments would flow. At the time of independence, the nation was faced with high unemployment, a poorly-educated and poorly-trained workforce, and a rapidly growing population. Being a labour-intensive economy then, Singapore embraced export-led industrialisation and promoted foreign direct investment. The focus was on low-tech manufacturing which was clearly illustrated by the government's emphasis on technical and industrial training for its population. This development strategy / model stood Singapore in good stead up to the early-1970s. However, it was the time to evolve as per changes that were taking place across the global stage. Other developing countries were gnawing into the low-tech manufacturing sector and giving Singapore tough competition in the sphere. Moreover, Singapore could not continue to maintain a labour-intensive model of industrialisation, as that was not viable owing to space constraints.

Singapore needed to respond to changing times and she did exactly that.

Science in Singapore in the Early Days

Back in 1965, Singapore had no compulsory education, and very few high schools and colleges of repute, resulting in a dearth of college graduates and skilled labour. The country was faced with a pressing need to build its economy and defence. To top it off, it had to deal with sustainability as most of its food, water and energy were imported. With a population of close to two million who were largely unskilled, the task at hand was a daunting one.

To match the rapid expansion of schools, a large scale recruitment of teachers was done. The result — universal primary education was attained in 1965 and universal lower secondary education by the early 1970s. More importantly, a public education system on a national scale was now in place. However, only 444 out of 1,000 students who joined in primary grade one reached secondary grade four after 10 years. Of this number, only 350 gained three or more passes in their O-level examinations. One report even estimated that between the years 1970 and 1975, Singapore would have a shortfall of 500 engineers and 1,000 technical workers, besides also being faced with a severe scarcity of people with management skills (Lee, et al., 2008).[4]

This was not good enough. Singapore had a lot of catching up to do, and particularly as she wanted to move from a low-skilled, labour-intensive economy to a higher-skilled economy higher up the value chain. This would require the creation of strong intellectual capital, which would only be possible through decisive governmental action and investment in science education, knowledge and innovation.

---

[4] OECD. *Singapore: Rapid Improvement Followed by Strong Performance.* 2010.

Petroleum and chemical products were the major exports in 1960, comprising 17.7% of the total manufactured exports. By 1970, the share of petroleum products alone increased to 38.1%. This was followed by electrical and electronic products that comprised 13.9%. Even foreign investments in mid-1971 followed a similar trend: petroleum industry (53%), metal, engineering, and transport equipment industries (10.5%), electrical-electronics industries (9%), and chemical industry (6%).[5]

Another major development in 1971 was the British withdrawal of forces from Singapore which meant that close to another 35,000 people were now unemployed. Creating employment for these newly unemployed people was another pressing need for the government, and part of the reason the government pushed for labour intensive industries such as construction, electronics, and shipbuilding and ship repairs.

By the mid-1970s, however, the focus shifted to capital intensive production and exports of (basic) technology-based products. These comprised precision engineering products, electrical/ electronic products, cameras, watches, photographic-optical products, ships and ship repair, rigs, and offshore drilling platforms. The shift towards greater technological sophistication was already being seen at this time.[6]

By the late-1970s, it was clear the government wanted to change the economic model from a labour-intensive to a capital- and skill-intensive one. With this goal in mind, a new education system was introduced in 1979 and upon this, the foundation of Singapore's future economic gains was to be based. This started with the birth of the Nanyang Technological Institute (NTI) in 1981 to produce more engineers for the country; NTI would soon morph into the Nanyang Technological University in 1991.

Regional competition started providing manufacturing facilities at cheaper rates than Singapore. To take the economy to the next higher level the government established the Agency for Science, Technology and Research (A*STAR) in 1991. Its mission statements states "A*STAR drives mission-oriented research that advances scientific discovery and technological innovation. We play a key role in nurturing and developing talent and leaders for our Research Institutes, the wider research community, and industry. Our research creates economic growth and jobs for Singapore..."[7]

In just three decades this transformation seemed unreal, but then, the world started to take notice of this tiny city-state Singapore. Philip Yeo, Chairman of

---

[5] Some basic problems of industrialisation in Singapore. Author: SY Lee. Journal of Developing Areas, Vol. 7, No. 2 (Jan., 1973), pp. 185–216.
[6] Ibid.
[7] Agency for Science, Technology and Research website.

A*STAR (2000–2007) summed it up neatly in an interview in 2005: "Our economy went through many different stages. We started in 1965 at high unemployment and worked ourselves up to full employment. We started with manufacturing industry, low-skill, labour-intensive, then steel and cotton industry, then chemical industry, then microchip and semi-conductor industry, then knowledge based industry. Knowledge is the key and the most important for knowledge is education, especially higher education."[8]

Creating Engineers

The development of Singapore's human resource, as said by the late Prime Minister Lee Kuan Yew would determine whether the nation would "sink or swim" (Minchin, 1990, p. 242).

In the 1980s, the Singapore Government adopted developmental strategies that would help the nation achieve an economy and society higher up the technological ladder. Several educational changes were introduced and thus began the phase which would be now termed the 'Second Industrial Revolution'. A Ten-Year Plan was laid out with an aim to take the manufacturing sector's share of GDP from 22% in 1979 to 31% in 1990. This required a restructuring in two areas: giving enough attractive incentives for the influx of MNCs to invest in high-technology operations, and promoting science and technology including activities like research and development.

To achieve this, the government once again revisited the education system. With the number of polytechnic and university graduates on the rise, the profile of the workforce changed. But the government realised that to be a technologically advanced nation, the country had to be able to produce its own local pool of scientists and engineers. The dearth of skilled research scientists and engineers was clear. In 1990, out of every 10,000 Singapore workers, 114 were engineers by qualification, but only 29 were research scientists and engineers. Once again, the government intervened and ensured that employment prospects and higher salaries were afforded to students from these streams. This was done through several public-private partnerships and the results are shown in the table below.[9] The output of science and engineering graduates had increased as envisioned.

---

[8] Re-inventing Society: State Concepts of Knowledge in Germany and Singapore. Author(s): Anna-Katharina HORNIDGE. Source: Sojourn: Journal of Social Issues in Southeast Asia, Vol. 22, No. 2 (October 2007), pp. 202–229. Published by: Institute of Southeast Asian Studies (ISEAS).

[9] World Bank. *The Development of Education in Singapore since 1965.* June 2006.

Output of Degree	1981–1989	
Arts	3792	4542
Science	3180	4105
Engineering	2467	5005

The discipline of Materials Science and Engineering started developing only during the early part of the 1980s, with the first Scanning Electron Microscope facility set up in the Singapore Institute of Science and Industrial Research (SISIR) in 1981. Mostly people were working on metals in what were usually called Departments of Metallurgy, but both NTU and NUS began hiring faculty in polymer science and composites from the mid 1980's.

Singapore's Primary Universities Come to the Fore

The first formal recognition of the importance of MSE came when NTU established the Division of Materials Engineering in 1991 to offer a 3-year programme for the Bachelor of Applied Science (Materials Engineering) degree, and an optional 4-year programme for the Bachelor of Applied Science with honours. The following year, a 4-year direct honours programme that led to the Bachelor of Engineering (Materials Engineering) was launched. The Division expanded into a School of Materials Engineering by itself in the year 2000, changing its name to Materials Science and Engineering in 2005. The National University of Singapore (NUS) also set up a Department of Materials Science in 1996 in the Faculty of Science, and awarded Bachelor of Applied Science Degrees. This later became the Department of Materials Science & Engineering within the Faculty of Engineering in 2005.

The timing of the launch of these programmes worked well for Singapore for at that time, the number of graduates being produced in this field was still small. Consequently, there was an insufficient number of graduates to fill the job openings in the local market. It was in this regard that the universities in Singapore played a critical role through the evolution of comprehensive MSE programmes.

In 1997, the government established the International Academic Advisory Panel (IAAP) to advise on the research direction and strategies which NUS and NTU should adopt in university education, based on world trends. It was also the IAAP's job to help NUS and NTU partner with leading universities and research institutions across the world and establish joint efforts in research, staff and student exchanges.

In the year 2000 when the Division of Materials Engineering at NTU was converted into a full-fledged School of Materials Engineering (SME) under the leadership of the late Professor Fong Hock Sun, 133 students completed their undergraduate programme in this field at NTU, up from just 21 in 1994. The school's name was changed to the School of Materials Science and Engineering in the year 2005 when its Dean, Freddy Boey, steered the school to focus strongly in graduating students with strong competencies in research. By 2006, the school grew to be the biggest MSE school or department in the world, exceeding a MSE student number of 1000 over the 4 years programme. Each year since then, about 250 undergraduates have convocated every year and gainfully got employed. Comparatively, the school was graduating more MSE engineers than all MSE departments combined in the UK. The remarkable success of the School was also seen in the PhD graduate numbers, which exceeded 200 since 2006. The remarkable achievements of the school extended to its innovation efforts. The school had its first start up company back in 2004 in the form of Amaranth Medical, funded by a Silicon Valley VC and having presence in both Mountain View Ca and Singapore. The company developed one of the world's first drug eluting heart stent and has since claimed Boston Scientific as one of its major investors. Since then, 13 companies have been spun off by its faculty and 7 companies have been started up by its graduates. As of January, 2015 the School has 200 research staff, 273 PhD and 875 undergraduate students. In the year 2004, the school registered 4 patents, and this number increased to 70 in 2011 and shot up to 96 in 2013 (see figure below).

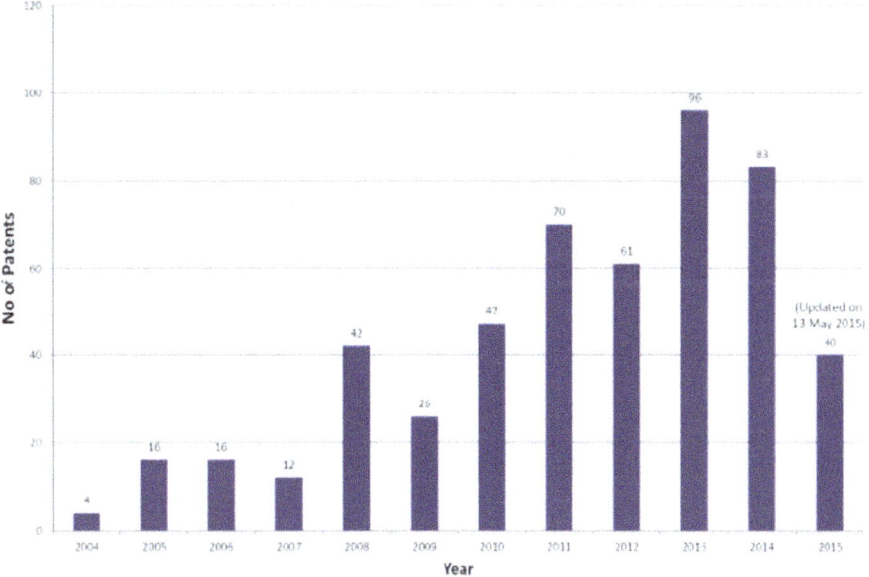

Figure. Number of patents registered by NTU's School of Materials Science and Engineering.

This demonstrates that not only had Singapore's education system produced a sufficient number of graduates to meet the needs of the economy, but it had also been producing researchers and research of the highest standard, putting Singapore firmly on the map as one of the premier MSE research centres around the world. The School of Materials Science and Engineering at NTU has done and continues to do research that is highly relevant and cited (see Figure below).

When the Chair of the School Prof Freddy Boey took up the Provost position in 2011, his Assoc Chair Professor Ma Jan was chosen to Chair the school. Most unfortunately, Prof Ma passed away prematurely in 2012 and was replaced by its current Chair, Prof Subbu Venkatraman. In 2013, NTU launched a fund in his memory to provide scholarships and bursaries for students at the school. NTU also renamed its High Speed Dynamics Laboratory at the university the Ma Jan High Speed Dynamics Laboratory in his honour for his accomplishments and his contributions to the school.

The School of Materials Science and Engineering at NTU has come a long way in a short span of time. It helped NTU to be the second highest ranked in the world in MSE by the US News in 2015, second only to MIT.

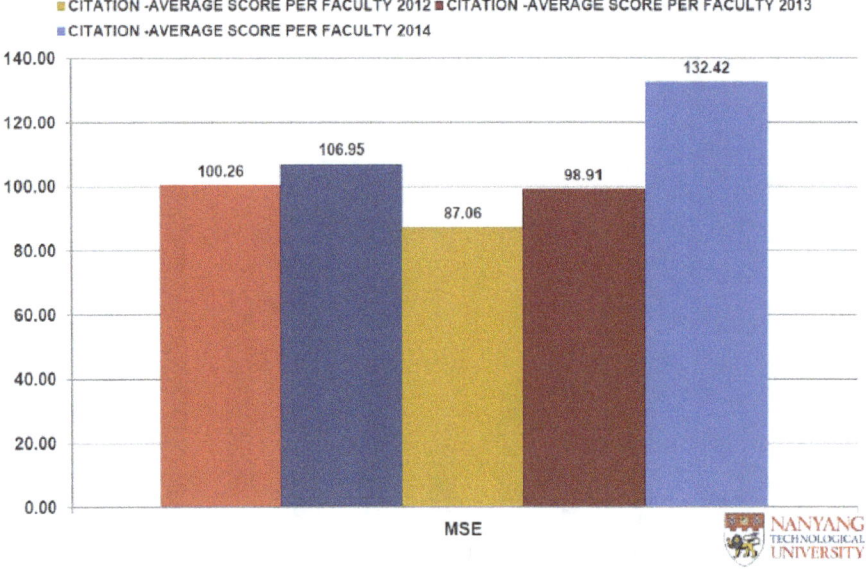

Figure. Citation Score per Faculty Member at the School of Materials Science and Engineering.

The Research Push

Singapore acknowledged early enough that while training enough graduates was essential in making Singapore a sophisticated manufacturing powerhouse, it was not going to be enough in helping Singapore stay ahead of the curve in this area. The government realised that fostering research and development in the area of materials science was going to be important for the country if it was to continue attracting global companies to invest in the country.

Research is the cornerstone of Singapore's increased prominence in this area of study. Increasingly companies from around the world have expressed interest in collaborating with our universities for the purposes of research. This is something we can be very proud of.

One of the leading research centres in materials science, the Institute of Materials Research and Engineering (IMRE) was established as a part of Singapore's Agency for Science, Technology and Research (A*STAR) in 1997. Having built strong capabilities in materials analysis, characterisation, design and growth, patterning, fabrication, synthesis and integration, the IMRE had partnered with institutes and the industry to create sites for world-class materials science research. Areas of R&D include research on organic solar cells, nano-composites, flexible organic light-emitting diodes (OLEDs), solid-state lighting, nanoimprinting, microfluidics and next generation atomic scale interconnect technology.

The IMRE helps in core competence in critical technology areas and enables fundamental new discoveries and the development of advanced materials that would eventually lead to new technologies and new commercial products. Through its collaborations and efforts, it is one of the leading research institutes for MSE globally.

Several new multi-million dollar research institutes had been launched in NTU with the support of government agencies such as the National Research Foundation, the Ministry of Education (MoE), the Economic Development Board (EDB), the Environment and Water Industry Development Council and the Media Development Authority (MDA).

With a strong focus then on sustainability given the prevailing environment concerns of our times, the Singapore Centre on Environmental Life Sciences Engineering, Nanyang Environment and Water Research Institute (NEWRI), and Energy Research Institute at NTU (ERI@N) had been established. These centres conducted research on aspects of environmental sustainability and the sustainability of the world we live in. Much of this research was centred on materials and what could be achieved through tweaking with them.

In particular, NEWRI and ERI@N have been of critical importance at both a local and global scale as they have been working on water remediation, environmental sustainability, clean energy generation, storage and management.

ERI@N, which was officially established in June 2010, aims to be a leading research institute for innovative energy solutions. Its mission was to be a centre-of-excellence for advanced research, development, and demonstration of innovative energy solutions with global impact. Research at ERI@N encompasses seven programmes, namely, fuel cells, energy storage, sustainable buildings technologies, solar cells and solar fuels, maritime clean energy, wind and marine renewables, and electro-mobility. Its two flagship projects, namely the EcoCampus and Renewable Energy Integration Demonstrator Singapore (REIDS), focus on significant outcomes in energy efficiency and renewables. The former, a partnership between NTU, Jurong Town Corporation and EDB, would transform the NTU campus and JTC's Clean Technology Park into the most sustainable college campus in the world and would reduce the energy utilisation on campus by at least 35% over the next decade.

REIDS on the other hand, would integrate solar, wind, marine, and bio-renewables on the island of Semakau, off the southern coast of Singapore. The facility would be an island micro-grid that provides an ideal environment to develop and export micro-grid solutions uniquely suited to tropical conditions in countries ranging from Africa, Central and South East Asia.

With 170 full-time staff and 138 PhD/ Master scholars, ERI@N had set up key collaborations with University of California, Berkeley, Technical University of Munich, University of Cambridge, and Lawrence Berkeley National Lab; as well as 33 industry partnership projects, including joint laboratories with global industry leaders such as BMW Group, Rolls-Royce, Johnson Matthey, and Vestas amongst others.

Emphasising its position in environment and water technologies (EWT), NTU launched the Nanyang Environment and Water Research Institute (NEWRI) in 2008. NEWRI was an 'eco-system' of research groups, centres and institutes which create a holistic value chain. It serves as a one-stop centre for industry and research partners to access NTU's EWT capabilities. To meet its vision, NEWRI has three facilities currently. They comprise the Environmental Chemistry and Materials Facility, the Environmental Biotechnology Facility, and the Membrane Technology Facility. NEWRI was recently ranked among the world's top 9 notable water research organisations. It has been able to create a platform for approximately 400 researchers to perform RTDA (Research, Translation, Development and Applications) in the domains of Water, Waste, Wastewater, and the Energy-water nexus to create technological solutions that can be applied to both community and industry.

Another important research area for NTU was defence science materials which has been substantially funded by the Defence Research and Technology Office and supported by the Defence Science Organization National Laboratories. The areas of work include novel graded functional materials, electrochromism, electro-optics and smart textiles. Temasek Laboratories@NTU (TL@NTU) was a partnership established in 2003 by the Defence Science & Technology Agency (DSTA) and NTU for the coordination and development of long term technology programmes. TL@NTU would look to explore the frontiers in science and develop strategic technology that would deliver effective solutions to enhance the defence and security of Singapore. It aims to carry out innovative R&D programmes that would lead to technological breakthrough in defence science and technology. TL@NTU would also establish partnership programmes and effective collaboration with the universities, defence community and industry to facilitate the transition of technology to applications, provide technical advice to our partners, and provide a forum for dialogue on defence technology and practices.

The Singapore government is anticipating that the next generation of materials would come from mimicking natural materials (this discipline is termed biomimetic materials). Examples include an artificial leaf concept for solar energy conversion; materials for hard and soft tissue repair; and superhydrophobic and anti-fouling coatings to name a few. To this end, a centre called the Centre for Integrative Sciences was set up by A*STAR.

In 2009, the Center for Biomimetic Sensor Science (CBSS) was established in partnership between NTU, Linköping University (LiU), Sweden and the Austrian Institute of Technology (AIT), Austria to drive a new research initiative on Biomimetic Sensing. The aim was to establish strong and truly multidisciplinary bio- and chemical sensor activity at NTU. Biomimetic concepts in conjunction with recent advances in biomolecular- and nano-sciences were identified as key components to exploit new sensor technology for biomedical, environmental and industrial applications. Besides the technology driven sensor projects, the centre also aimed at developing a deeper fundamental understanding of optical and electrical transduction mechanisms; and structure-function relationships of potential soft matter sensor materials/architectures. Thus, CBSS was pursuing both fundamental and applied research projects. Another important mission was to establish strong links to industrial partners active in the diagnostic and (bio) analytical sectors. Today, CBSS hosts about 45 students and research staff.

Driving research in partnership with the industry, NTU launched an S$5.3 million joint lab with British chemicals giant, Johnson Matthey in 2014. NTU would be carrying out joint research in new materials and renewable energy solutions with the multinational company, for whom it was the first research collaboration of this kind in Asia. A leader in sustainable technologies with more than 11,000 employees worldwide, Johnson Matthey's partnership with NTU would focus on the three key areas of energy storage, air purification and membranes, and energy and catalysis. Over the next four years, the new Johnson Matthey @ NTU lab aims to develop breakthroughs for battery technologies, better air purification systems, and valuable chemicals made from biomass and other related fields.

Ocular Therapeutic Engineering Centre (OTEC) was founded in 2011 by Professor Tina Wong from Singapore Eye Research Institute and Professor Subbu Venkatraman. The mission of OTEC was to provide safer, more efficacious and patient-friendly therapeutic options through scientific breakthroughs, cutting edge and innovative technology in MSE. OTEC would be where material scientists and engineers meet clinical researchers to come up with novel solutions to address an unmet clinical or surgical need in ophthalmology, with the ultimate goal to enhance existing patient care and improve on our current clinical management outcomes. Current research activities focus on the development of sustained-release drug delivery systems, the use of sophisticated materials to design and engineer improved surgical devices, and the development of faster, more accurate imaging devices for screening and diagnostic purposes in ophthalmology. Currently, OTEC has 10 research scientists and two PhD students as well as several attachment students from NTU with a research funding at S$4.5 million.

Drugs have long been used to improve health and extend lives. Drug delivery systems are engineered technologies for the targeted delivery and/or controlled release of therapeutic agents. The practice of drug delivery has changed dramatically in the last few decades and even greater changes are anticipated in the near future. Biomedical engineers have not only contributed substantially to our understanding of the physiological barriers to efficient drug delivery — such as transport in the circulatory system and drug movement through cells and tissues — they have contributed to the development of a number of new modes of drug delivery that have entered clinical practice. Drug delivery systems control the rate at which a drug is released and the location in the body where it is released. Some systems can control both. The Advanced Drug Delivery Centre (ADDEC) would look into addressing this vision and pioneer initiatives to lead an effort inclusive in the design of a wide variety of drug-delivery systems, surgical implants, artificial organs, and wound-closure devices that are critically dependent on biomaterials. The centre would also serve as a wellspring of ideas for knowledge creation, manpower training, technology transfer, and commercialisation. The main Singapore collaborators would be NTU's School of Materials Science and Engineering and LKC School of Medicine at NTU coupled with Mayo Clinic. In order to train future research personnel, develop knowledge, IP and intellectual credibility, ADDEC would also look into supporting a joint postgraduate programme for doctorate and master's degrees through NTU and LKC School of Medicine.

The Singapore-Berkeley Research Initiative for Sustainable Energy (SinBeRISE) programme was another initiative for research that involves materials science.

Established in 2012, this was a research programme funded by the National Research Foundation (NRF) under its Campus for Research Excellence and Technological Enterprise (CREATE) programme. This was the second joint research programme between University of California, Berkeley (UCB), NTU and NUS.

The first programme, SinBerBEST, was started in January 2012. Research SinBeRISE seeks to harness solar energy using novel, inexpensive approaches. It aimed to substantially improve the overall efficiency of photovoltaic (PV) devices and harvest solar energy for conversion into electricity and fuels. The programme comprises three main areas: direct conversion of solar energy to electrical energy; generation of clean fuel using solar energy; and the integration of the first two areas into functional systems and devices.

The Technion CREATE Programme was launched in 2010 and funded by the National Research Foundation (NRF). Through this partnership, NTU, NUS and the Israel Institute of Technology (Technion), aimed to address the clinical need for cardiac restoration therapy using a tissue-engineering based approach. The programme's primary focus was the development of a functional tissue-engineered bioactive cardiac patch. The use of a fully functional cardiac patch that can be surgically attached and integrated onto the MI region of the injured heart was seen as having great potential to restore cardiac function effectively. To date, the project has successfully refined our porcine cardiac extracellular matrix technology to produce thick acellular scaffolds with inherent vasculature that resemble the native tissue, and can support the long term survival of several progenitor cardiac cells.

In July 2011, a joint research programme between NTU, Hebrew University of Jerusalem (HUJ), and Ben Gurion University (BGU) was started on Nanomaterials for Energy and Water Management. The NEW CREATE 5-year collaborative research programme aims to address the need for efficient technological management of both Energy and Water by developing specially tailored Nanomaterials. Apart from achieving its scientific and technological objectives, the programme would also promote the emerging energy and water industries in Israel and Singapore by establishing start-up companies and licensing technologies to existing companies. A nucleus of scientists would create a state-of-the-art center using nanomaterials in technologies which have high societal impact.

A total of 79 researchers have been working in this programme including post-doctoral fellows, research associates, project officers and PhD students. The research team comprises 10 Principal Investigators, 4 from NTU, 3 from the Hebrew University of Jerusalem and 4 from the Ben Gurion University of the Negev. The combined research effort has generated over 30 patents.

Since 2014, NTU has also been collaborating with the International Institute of Nanotechnology (IIN) based in Northwestern University to establish the NTU-

Northwestern Institute for Nanomedicine (NNIN) with a $32M funding. Nanomedicine is a relatively new discipline that combines nanomaterials-based detection and treatment of diseases and is expected to have the most profound impact and benefit to society. The institute would bring together scientists from both universities to collaborate on research projects in the areas of disease diagnostics, timed-release therapeutics, and targeted drug delivery methods. NNIN would also focus on developing innovative solutions in four key areas: diabetes, cardiovascular, ophthalmology and skin therapeutics. The institute would build upon the university's success in nanomaterials to explore innovative solutions to medical problems using approaches that use nanotechnology. NNIN embodied an extraordinary combination of scientific capabilities, education programmes, and partnerships that provided a unique platform from which students, engineers, scientists and clinicians could work together and address medical needs which are currently not addressed.

The Singapore-MIT Alliance for Research and Technology (SMART) was another leading research centre that incorporates materials. It was established by the Massachusetts Institute of Technology (MIT) and the National Research Foundation (NRF) of Singapore. Its core objective was to identify and conduct research on critical problems of societal significance for today and the future. SMART's five Interdisciplinary Research Groups comprise Infectious Diseases, Center for Environmental Sensing and Modeling, BioSystems and Micromechanics, Future Urban Mobility, and Low Energy Electronic Systems.[10]

There have been various partnerships with the private sector as well, furthering research and learning in this area. Singapore partners with eminent global players of the industry in the pursuit of innovative materials research.

Vestas, a global leader in providing wind energy solutions has set up a joint laboratory with NTU to focus on materials research for wind turbine applications, while Gamesa has set up its first advanced materials research centre in Southeast Asia at NTU. British engineering giant Rolls-Royce has been working with NTU in applied engineering and technology applications. Bosch has also teamed up with ERI@N to develop solutions in organic photovoltaics, a cost effective alternative to silicon-based solar cells. Australian renewable energy firm Dyesol has recently forged a partnership with ERI@N to research on and develop low-cost dye sensitised solar cells technology.

These research collaborations/ partnerships culminate in benefits for both Singapore as well as the corporations themselves.

Let's take the Rolls Royce case for an example. Rolls Royce started working with Singapore, in particular with NTU, more than 10 years ago and started setting up a big plant. Rolls Royce's new generation engines are now developed and manufac-

---

[10] Campus for Research Excellence and Technological Enterprise website.

tured in Singapore. First, the company needed to be convinced about the quality of engineers before hiring. Trained engineers were readily available in Singapore — both NTU and NUS were producing many highly skilled engineers. Second, a very high level of competency in MSE and related research was required as the company's operations had to do with turbine plates and developing new generation turbine plates in Singapore. Singapore was and is producing a lot of MSE related doctorate graduates. Because of this availability of talent, Rolls Royce decided to setup shop in Singapore.

Singapore is a key centre for Rolls-Royce. Today, the company has its most modern research and manufacturing facility in Singapore with more than 2,200 employees, of which 90% are local — most of whom are graduates from Singaporean technical institutes. As of end-2013, Rolls Royce accounted for more than 15% of the country's aerospace output. Additionally, in the defence aerospace sector, Rolls-Royce is a major supplier to the Singapore Air Force. It was estimated that the company's activities in Singapore would have directly and indirectly contributed approximately 0.5% to Singapore's GDP by the end of 2015.

Over time, the School of Materials Science and Engineering at NTU has continued to remain at the forefront of several external partnerships, resulting in research funding being provided for the school. The external annualised research funding for MSE is shown below:

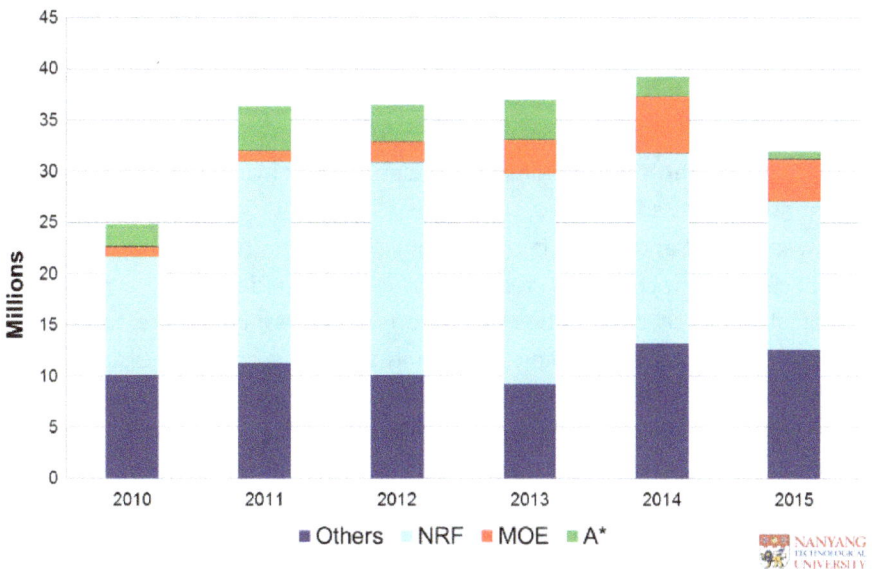

Figure. External Annualised Research Funding for the School of Materials Science and Engineering at NTU.

The MSE fraternity in Singapore has benefited as much from the industry as it has from the government. Both sides have provided a good amount of funding and this is a sustainable model. MSE has become the key driver in so many industries like energy, water, environment, biomedics — and these are just a few. The future of MSE and research in the field is not only looking up but delivering results, and changing the way industries and individuals are — being able to do more by using less. And as has been stated earlier, 90% of the material systems are still waiting to be discovered or invented. In short, it's like a golden sunset that never ends.

What we see in Singapore is a very rich MSE ecosystem with several pillars that are supporting one another. The government is clearly focused on making advances in the area of MSE as it realises the importance of the area of study for the future development of Singapore. The country's two leading science and technology universities have also realised the importance of MSE and the role of MSE graduates in the economy. Research centres — which were often set up in partnership between different educational institutions or between educational institutions and global companies — complement the work that was being done at the universities. This ecosystem ensures that budding engineers and scientists do not gravitate away from materials science and in so there would be sufficient funding to support the academic and research interests of young engineers and scientists working in this area.

Materials become a Cornerstone of Singapore's Economy

At the turn of the millennium, Singapore started focusing on biomedical sciences on a national scale. The vision of the government was telling as it not only identified and promoted the biomedical science sector but significantly funded it and continues to do so till date.

Between 2000 and 2005, it started building up its human, intellectual, and industrial capital that created a stepping stone and foundation for strong scientific biomedical research capabilities. Between 2006 and 2010, the focus was to strengthen its capabilities in clinical research and bring scientific discoveries that could be used beneficially on patients. The objective was to improve health care delivery systems that benefit the individual and the society at large, and eventually contribute to the economy.

Starting in 2001, investment in the biomedical sector has steadily increased. S$13.5 billion was committed by the government for R&D from 2006–2010 which was more than double the amount spent in the preceding five years. 25.3% of this amount was allocated to the biomedical sector. The gross expenditure on R&D

(GERD) as a part of the GDP grew at a compound annual rate of 11% from 2000 to 2008, and in 2008 GERD reached S$7.1 billion, or 2.8% of the GDP. The numbers were a clear indication of the rapid increase in R&D activities in Singapore. On recommendations by the Economic Strategies Committee, it was decided to target GERD at 3.5% of GDP by 2015 estimated at S$16.1 billion.

When things started looking bright, the global recession post 2008 came along plunging economies, cutting jobs and dissolving giant institutions. Even through this turmoil, Singapore stood steadfast in its commitment to R&D. This was best told by Prime Minister Lee Hsien Loong when he outlined the government's strategy for R&D in general and A*STAR in particular in 2009:

> *"We must expect slower growth and greater uncertainties at least over the next year, maybe longer. But our R&D programme takes a longer term perspective. It will proceed despite these immediate ups and downs. Its funding will not be affected. The Government remains fully committed to investing in R&D, in order to develop a key capability that will keep our economy competitive in the long term."* He added, *"Our steady commitment will continue to draw researchers to set up and root their research activities in Singapore, and give investors the confidence to establish high tech industries and corporate R&D centres here."*[11]

MSE leads the efforts in translation of biomedical products in Singapore. Already several companies have been founded and are doing well. However, the time horizon for the translation in biomedical products is generally long, but MSE has shortened this timeframe in the nanomedicine area to about five years from concept to clinic. Several companies such as Amaranth Medical, Medlinx Acacia, Peregrine Ophthalmic, ACM Biolabs, Osteopore were all start ups founded by MSE IPs in the university.

Today, Singapore houses more than 100 global biomedical sciences companies operating in several business areas. Companies such AB SCIEX, Baxter International, Becton Dickinson, BIOTRONIK, GlaxoSmithKline, Hoya Surgical Optics, Life Technologies, Medtronic, Novartis, Siemens Medical Instruments, Takeda, Abbott, Roche, Merck, Pfizer, Schering-Plough, and Wyeth are just a few of the big names that have a presence in Singapore.

How does this relate to in terms of economics? The manufacturing output for biomedical sciences increased from S$6.3 billion in 2000 to S$23.3 billion in 2010, accounting for about 5% of the country's GDP. In the same period, the number of

---

[11] Agency for Science, Technology and Research. *Singapore opens Fusionopolis, its second major R&D hub in 5 years.* Available online.

people it employs has more than doubled from 6,000 to 14,000 and the number of people hired in biomedical research and development has increased from 2,200 to 5,000. Finally, as of 2013, the biomedical sciences sector was the second largest in the manufacturing segment of Singapore's economy, contributing 20.5% of the country's total manufacturing output.[12,13] The manufacturing output is estimated to hit S$25 billion by 2015, targeting to employ 15,000 people by the year 2015.

In the area of biomaterials, Amaranth Medical, founded by Professors Freddy Boey and Subbu Venkatraman, has more than 150 human patients with their fully-degradable stents in a FDA multi site clinical trial. Both professors have also started Peregrine, which has been developing drug eluting liposome based injectable nanocarriers for treating glaucoma. The school also holds the record of having one of the first biomedical device approved by FDA for sale, through the start up Medlinx Acacia, in the form of a customisable PVDF hernia mesh. ACM Biolabs, another start up from the school, has been already marketing products related to biomarkers of diseases.

In the area of electronic materials, a super-capacitor based electronic flash has been co-developed by our faculty and the company Xenon. Printed Power, founded by Prof Subodh Mhaisalkar, has been developing printed batteries and sensor networks. A concept of superfast charging of LI-ion batteries has now been spun off as Quick Charge; a spin-off company called Nanosensormatics has been developing nanosensors using nanostructures. In the area of nanomaterials, we have a spin-off company called Salus Nanotechnologies which produces silver nano particles for coating on end products. The silver based nanomaterials have wide applications in the medical, cosmetic, water purification, and air disinfection industry.

In the energy sector, a new perovskite material for solar energy conversion with the potential for significant improvement over silicon has been discovered by our researchers. A spin-off company called Amperics provides cost effective, high energy and power density energy storage solutions for the Uninterrupted Power Supply (UPS), light electric vehicles, and hand-held power tools markets. This solution based on carbon nanomaterials and low-cost transition metal oxides provides a cost effective and scalable alternative with applicability to the broader electromobility markets.

At NUS, the graphene centre has been developing 2-dimensional materials for use in aerospace, energy and biomedical sectors. NUS's Graphene Research Centre (GRC) and the world's leading chemical company BASF have joined hands

---

[12] Belgian Foreign Trade Agency. *Biomedical Science Sector: Singapore*. July 2014.
[13] Issues in Science and Technology. *Innovation Policy Around the World: Singapore Betting on Biomedical Science*. Lim Chuan Poh. 2010.

to develop the use of graphene in organic electronic devices, such as organic light emitting diodes (OLED). The goal of this collaboration has been to interface graphene films with organic electronic materials to create more efficient and more flexible lighting devices.

These are but a few examples of areas in which materials science has been making a real difference in Singapore. Today, almost all kinds of manufacturing industries are in some way or the other affected by materials science. This ranges from oil and gas and biomedical science to basic household products; and from electronics for both commercial and personal use to water and waste management. Any industry that uses any kind of material has been affected and influenced by advances made in the field of materials science.

The Presence of a Complete Ecosystem

MSE has become an entrenched discipline in Singapore with a large critical mass of faculty members, engineers, researchers, students and MSE related private companies. There has been a thriving ecosystem in place in Singapore where we continue to see research collaborations between corporates and our local universities, and between government-funded research centres and the universities. The government has been supportive of the MSE sector with funding being provided to all the research initiatives and that has encouraged the development of the sector too.

As a result of these initiatives of our government, the universities and research centres, Singapore has come to be known as a meeting centre for leading materials scientists globally, including some of the top minds in the area of study. Much of this has been made possible by Materials Research Society of Singapore (MRS-S), and the efforts have helped in truly putting Singapore on the MSE map globally. Singapore is home to one of the largest conferences in the MSE space. The International Conference on Materials for Advanced Technologies (ICMAT) is a biennial conference which attracts more than 2,000 delegates from around the world. Organised by the MRS-S, this conference is a major event that sees a line-up of Nobel Laureates, distinguished professors and researchers from all over the world.

These are, in short, sectors that we expect to drive Singapore's economic development over the next decade or two. Our focus on materials is in keeping with the aspirations of Singapore's economic development. The future looks good, with more research companies setting up facilities in Singapore. There have been also an increasing number of home-grown companies that would survive and grow into a reasonably big size and hire more people over the years.

One thing is clear, and that is Singapore has consistently evolved. The *Economist* magazine summarises Singapore's vision best: "In the 1980s, thanks to Seagate, an

American firm, and other multinationals, Singapore accounted for 60% of hard-disk drives (HDDs) shipped globally. As production moved to Thailand, Malaysia and China in the 1990s, Singapore became the centre for production of higher-margin "enterprise HDDs". By the early 2000s Singapore had 80% of this global market and had already begun to shift to the next level, hard-disk media, in which it now has a market share of about 40%."[14] This example illustrates Singapore's understanding of industry, and its government's light footed nimbleness whenever it is faced with a need to change its economic structure or objectives. Singapore has been this way for the last 50 years, and continues to exhibit a sharp understanding of the global economic landscape and its position in that context.

At this point, in 2015, the government has announced that it would focus on five growth clusters: advanced manufacturing, aerospace and logistics, applied health sciences, smart urban solutions, and financial services. Of these, MSE is the cornerstone of the first four, demonstrating the importance of this field of study to Singapore and Singapore's continued prosperity.

Singapore has carved a name for itself globally in the area of MSE. Both NTU and NUS have been globally recognized as doing world class research in MSE and graduating sufficient number of capable, skilled engineers and scientists in this area every year. The future is looking bright, and so it should be, for materials science would continue to play a large role in the future of Singapore in many areas. Of one thing we can be certain — that Singapore will continue taking the steps necessary to enhance its position as a leading global centre in materials research and engineering.

---

[14] The Economist. *Many spokes to its hub*. 18 July 2015.

2 Composites, Nanocomposites and Hybrid Materials

Chaobin He[*], Xiao Hu[†], Zhang Yu[*] and John Wang[‡]

The research into composite, nanocomposites and hybrid materials in Singapore started in the later 1980s, mainly among the research community at the National University of Singapore (NUS), Nanyang Technological University (NTU) and later in the A*STAR research institutes. The works then were mostly focused on the processing, the mechanics and the non-destructive testing of conventional glass fibre and carbon fibre composites. The important factors that propelled the composites, nanocomposites and hybrid materials research in Singapore were the substantial increase of research funding support by the government to the local universities and the establishment of the Institute of Materials Research and Engineering (IMRE) at A*STAR in the late 1990s, as well the local industrial pull which demanded sophisticated materials to support their product development. Now Singapore has one of the world's largest materials engineering community equipped with sound infrastructure, state-of-the-art facilities including those for composites and nanocomposite research. The research into advanced composites research in Singapore has experienced an unparalleled rise in output and impact. This review will focus on the applications of advanced composites as mechanical, thermal and electrical conductive materials. The applications in the biomedical, energy, and environmental areas will be discussed in other chapters.

Table 1 shows the statistics of the research output in terms of academic papers and their accumulated citations from 1991 to 2015. Within the short span of less than 25 years, Singapore's number of scientific and engineering journal publications in the field of composites and hybrid materials have grown impressively from 196

---

[*] Institute of Materials Research & Engineering (IMRE), Singapore.
[†] School of Materials Science & Engineering, Nanyang Technological University.
[‡] Department of Materials Science and Engineering, National University of Singapore.

Table 1. Journal Papers Published by Researchers in Singapore in the Field of Composites, Nanocomposites and Hybrid Materials and their Citations.

Year	Number of Papers	Times Cited*
1991–1995	196	3051
1996–2000	791	13100
2001–2005	1432	44819
2006–2010	1742	57043
2011–2015	3164	47593

*Citation Counts are Cumulating from the Date the Papers were Published up to the Present. SCI (Science Citation Index)[1]

(1991–1995) to 3164 (2011–2015) by 1610%, and their citations rose dramatically from 3051 to 47593 by 1560%.[1]

The advanced composites and hybrid materials research in Singapore has worked towards driving and enabling applications in several key industrial sectors in Singapore. Over the years, the research community in Singapore led and participated in a number of research and development programmes with industrial partners in various sectors including chemicals, microelectronics, aerospace, healthcare, defence, oil & gas, off-shore, energy and environment. Composites, nanocomposite and hybrid materials are polymers incorporated with inorganic fillers. Inorganic filler has long been used to prepare polymer nanocomposites and nanohybrids. The incorporation of nanomaterials can improve a polymer's properties such as its tensile strength, Young's modulus, impact and scratch resistance, electrical and thermal conductivity, thermal stability and fire resistance. In the past two decades, there have been emerging efforts on the development of polymeric nanocomposites, where at least one of the dimensions of the filler material is of the order of nanometer scales. Polymer nanocomposites are manufactured commercially for diverse applications such as sporting goods, aerospace components, automobiles, etc. On the other hand, supramolecular nanohybrids of organic and inorganic components are also promising materials. These nanohybrids may exhibit unique properties such as amphiphility and thermal responsivity in a solution environment and the properties could be well tailored by molecular design. Organic-inorganic nanohybrids have diverse applications in drug delivery, cosmetics, viscosity and surface modification, and water purification.

The academic output in the field of composites and hybrid materials shown in Table 1 is astonishing by any standard and it is impossible to detail all aspects of the research in this commemorative chapter. Therefore, only a few examples are mentioned here.

Polymeric Nanocomposites and Applications

The research group at NTU advanced polymer resin and composite development has served the needs for defence innovation and capability expansion. In particular, the Advanced Functional Materials Programme at the Temasek Laboratories in NTU (TL@NTU) launched several research projects on composite materials working towards enabling technology and capability development. Several composite technologies developed under this programme have been transferred to the relevant industry and/or agencies.[2] Another example worth mentioning is the use of composites and hybrid materials in water treatment and environment protection. Among many other achievements, NTU's recent effort of using hybrid materials for water treatment has made a breakthrough in the development of a new concept for the point-of-use water purification and disinfection devices. The work has received the Best Paper Award in the technology category from a leading journal Environmental Science and Technology. The work was also highlighted by the top scientific journal NATURE and other scientific magazines.[3–6]

The technology, with pending patent, by NTU is able to disinfect water from various contaminated water sources into drinking water within several minutes. This new, safe and highly effective water disinfection method was developed based on polymer cryogel composites endowed with antimicrobial function using hybrid nanomaterials. The technology is targeted for multiple life-saving applications including for disaster relief in the aftermath of natural disasters and also as part of the emergency preparedness kit for outdoor adventurers and any other personnel who are exposed to the wilderness.

The key challenges in polymer nanocomposite research are twofold: the first is to develop a cost-effective approach to disperse/exfoliate fillers in polymer matrices, and the second is to elucidate the relationship between the processing, morphology and properties of the material. The research groups at IMRE and NUS have addressed the above two key challenges and successfully developed various nanocomposite systems, and some of the technologies developed have been patented and licensed.

In the work of Huang, et al.,[7,8] multi-walled carbon nanotubes (MWNTs) were treated in an acid environment to increase the content of carboxylic and hydroxyl groups on MWNTs' surface and this facilitates the dispersion of MWNTs in nylon-11. The incorporation of MWNTs significantly improved the thermal stability and enhanced the storage modulus of the polymer matrix. Liu, et al.[9,10] utilized a similar method to fabricate MWNTs/nylon-6 (PA6) nanocomposites. The elastic modulus and the yield strength of the composite are greatly improved by about 214% and 162% respectively, with incorporating only 2 wt % MWNTs. MWNTs/poly (methyl methacrylate) (PMMA) composites have also been fabricated by melt blending.[11,12] MWNTs were well dispersed in a PMMA matrix with no apparent breakage and the storage modulus of the PMMA is significantly increased by the incorporation of MWNTs particularly at high temperatures. Wang, et al. prepared biopolymer chitosan/MWNTs nanocomposites by a simple solution-evaporation method.[13] The MWNTs were homogeneously dispersed throughout the chitosan matrix. Compared with neat chitosan, mechanical properties, including the tensile modulus and strength of the nanocomposites, were greatly improved by about 93% and 99% respectively, with only 0.8 wt % of MWNTs in the chitosan matrix.

In the work of Shen, et al.,[14] organoclay, which contains functional groups on the basal plain, was used to replace pristine clay to reinforce epoxy by melt compounding. It was found that the incorporation of 7.5 wt % of organoclay enhanced the elastic modulus and hardness of the epoxy matrix by about 20% and 6% respectively. Organoclay was also well dispersed in chitosan,[15] nylon-6[16–18] and nylon-12[19] to effectively enhance the thermal and mechanical properties of the polymer matrix. To make direct use of pristine clay in the polymer nanocomposite fabrication, a scalable approach, i.e., "slurry compounding", to facilitate the exfoliation of pristine clay in epoxy was developed.[20–24] The "slurry compounding" method is illustrated in Fig. 1. In this process, the amount of organic modifier used is only 3–5 wt % of clay. The resulting epoxy/nanoclay composites exhibit a high degree

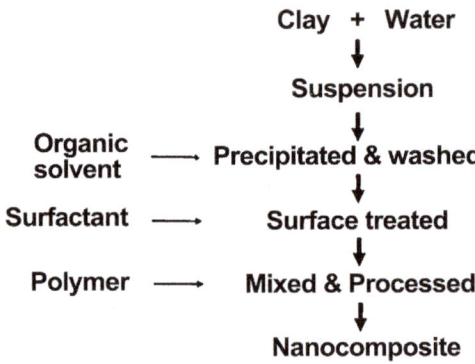

Fig. 1. Illustration of slurry compounding method.

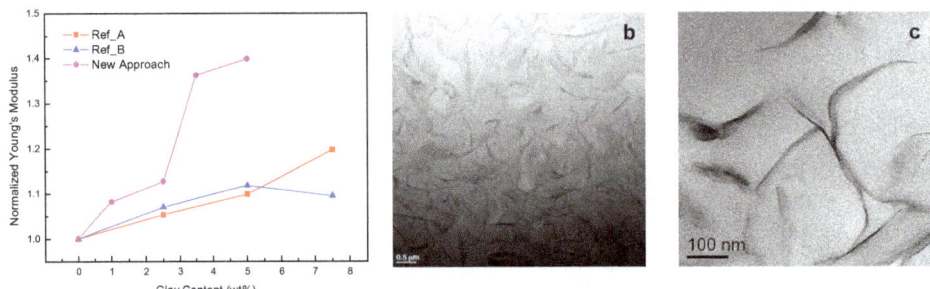

Fig. 2. Left: Mechanical property of polymer nanocomposite developed compared with other reported data. Right: TEM micrographs of polymer clay nanocomposite 2.5 wt %. Ref A and B are data from organic clay.

of clay exfoliation and a better thermal mechanical property. The mechanisms behind the clay exfoliation and the toughening effect were also elucidated and the technology has been exploited for many practical applications such as barrier film for food packaging and structural materials for transportation and marine and offshore industries, which resulted in patents filed and technologies licensing to companies.

For example, Wang, et al. fabricated epoxy/clay nanocomposites by the "slurry-compounding" process.[24] It was found that clay was highly exfoliated and uniformly dispersed in the resulting nanocomposites. Characterizations of mechanical and fracture behaviors revealed that Young's modulus increases monotonically with increasing clay concentrations while the fracture toughness shows a maximum at 2.5 wt % of clay. The initiation and development of microcracks are the dominant microdeformation and fracture mechanisms in the epoxy/clay nanocomposites. Most of the microcracks initiate between clay layers. The formation of a large number of microcracks and the increase in the fracture surface area due to crack deflection are the major toughening mechanisms. The "slurry compounding" method was also applied to a polyimide system,[25] which showed increased decomposition temperature and storage modulus with increasing clay content.

Another type of layered fillers, graphene, has been an important research topic in recent years, which includes the application of graphene in polymer composite fabrication. Bao, et al. fabricated a freestanding membrane composed of a nanofiber network of a graphene–polymer nanocomposite by electrospinning, which could be applied as an optical element in fiber lasers.[26] A small loading (0.07 wt %) of functionalized graphene enhances the total optical absorption of poly(vinyl acetate) (PVAc) by 10 times. The electrospun graphene–polymer nanocomposites exhibit wideband saturable absorbance for laser pulse shaping, and attain a larger modulation depth and smaller nonsaturable loss than single-walled carbon nanotubes.

Although desirable dispersion of inorganic fillers in a polymer matrix was achieved by the above simple surface modification and slurry compounding, the filler/matrix interactions are still not strong enough to provide efficient load transfer due to the lack of a strong bonding between the fillers and the matrix. On the other hand, strong filler/matrix interactions may further promote the dispersion of inorganic fillers. Hence, strong bonding, e.g., covalent bonds, may be necessary in polymer nanocomposite fabrication. In the work of Sin, et al.,[27] polyurethane (PU)/clay nanocomposites with and without clay-PU covalent bonds were comparatively studied. It was found that the clay-PU covalent bonds were favorable of load transfer between clay and PU and the resulting nanocomposites showed substantially higher Young's modulus and tensile strength. Various nanocomposites with strong covalent filler/matrix interactions have been studied. Pramoda, et al. initiated the polymerization of methyl methacrylate (MMA) from graphene oxide to form

PMMA-graphene nanocomposites.[28] With only 0.5 wt % graphene nanosheets, the T_g increased from 119 °C for neat PMMA to 131 °C for PMMA-graphene nanocomposite, and the respective storage modulus increased from 1.29 to 2 GPa. In another study,[29] polyoctahedral silsesquioxanes (POSS) molecules were grafted to polyimide (PI) chains through amine bonds and the PI-POSS nanocomposites showed much enhanced Young's modulus and tensile strength. In the work of Huang, et al.,[30] a series of functional POSS/PI nanocomposites were prepared, where there were strong covalent bonds between POSS and PI. The well-defined "hard particles" (POSS) and the strong covalent bonds between PI and the "hard particles" led to a significant improvement in the thermal mechanical properties of the resulting nanocomposites. The glass transition temperature dramatically increased while the coefficient of thermal expansion (CTE) decreased, owing to the significant increase of crosslinking density in the PI–POSS nanocomposites.

Strength and toughness are the two most important prerequisites for materials applications. Unfortunately, these two properties are often in conflict in materials. Yang, et al. reported the mechanical reinforcement of polyethylene (PE) using polyethylene-grafted multiwalled carbon nanotubes (PE-g-MWNTs).[31] In their research, the stiffness, strength, ductility and toughness of PE are all improved by the addition of PE-g-MWNTs. However, there was no clear explanation of the mechanisms behind the excellent mechanical enhancement. Recently, the research group at NUS and IMRE also developed novel nano-fillers, which could simultaneously enhance the strength and toughness of composite materials. The outcome of the research could provide a solution to overcome the conflict between strength and toughness which often exists in composite materials.[32–35] Sun, et al. tried to tackle the conflict between strength and toughness by designing a reinforcing unit of rigid-rubber-matrix as

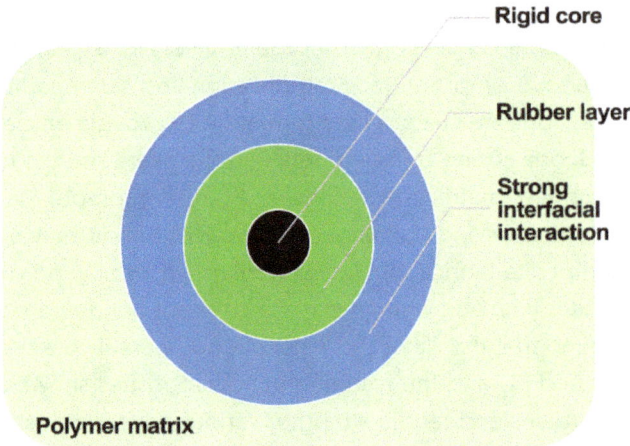

Fig. 3. Rigid-rubber-matrix strategy to simultaneously enhance strength and toughness of polymers.

shown in Fig. 3. There must be a strong bonding between the rigid filler and the rubber phase, which is responsible for the synergy of rigid and rubber phases. There must also be a strong bonding between the rubber phase and the polymer matrix and this is essential to effective load transfer. As an example, POSS was used to initiate the polymerization of the poly (caprolactone-co-lactide) (PCLLA) rubber phase, followed by the polymerization of poly (D-lactide).[32] The synthesized POSS-PCLLA-PDLA was blended with commercial poly (L-lactide) (PLLA) to form nanocomposites, where there is strong stereocomplex interaction between PDLA and PLLA. The resulting nanocomposites had a higher Young's modulus and tensile strength, and meanwhile exhibited increased elongation at break compared with neat PLLA. Lignin-PCLLA-PDLA was also useful in overcoming the conflict of strength and toughness of PLLA.[33] Besides, the new strategy was further applied to epoxy.[34] As shown in Fig. 4, Warintorn, et al. synthesized POSS-rubber-epoxy nanocomposites, whose Young's modulus, tensile strength, impact strength, elongation at break and fracture toughness were simultaneously enhanced compared with neat epoxy.

Polymeric composites have many applications, among which the biomedical field is an important example. In the work of Fujihara, et al.,[36] Guided bone regeneration (GBR) membranes were fabricated by electrospinning poly (caprolactone) (PCL)/CaCO$_3$ composite nanofibers on PCL substrates. In osteoblast attachment and proliferation experiments, absorbance intensity and tissue culture polystyrene of the composite GBR membranes increased during a 5-day seeding time. Ramakrishna, et al. fabricated poly (L-lactide) (PLLA) nanocomposite polymeric nanofibers containing nanohydroxyapatite (HA) by electrospinning.[37] Three nanofibrous scaffolds were fabricated, including PLLA, PLLA/HA and PLLA/collagen/HA. Osteoblasts were found to adhere and grow actively on PLLA/collagen/HA nanofibers with enhanced mineral deposition of 57%, higher than that of the PLLA/HA nanofibers. The synergistic effect of the presence of an extracellular matrix protein, collagen and HA in PLLA/collagen/HA nanofibers provided cell recognition sites together with apatite for cell proliferation and osteoconduction necessary for mineralization and

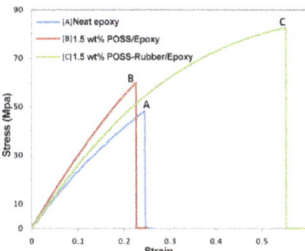
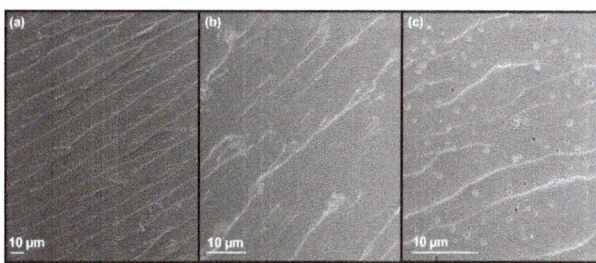

Fig. 4. Left: Stress-strain curves of neat epoxy and epoxy nanocomposites with 1.5 wt % filler. Right: Fracture surface morphology of (a) neat epoxy, (b) POSS/Epoxy and (c) POSS-Rubber/Epoxy nanocomposite at 1.5 wt % filler.

bone formation, indicating that the PLLA/collagen/HA nanofibrous scaffold could be a potential substrate for bone regeneration. They also reported nanofibrous scaffolds of poly(L-lactic acid)-co-poly(ε-caprolactone), gelatin and HA,[38] which is superior and suitable for bone tissue regeneration.

Organic-inorganic Hybrids and Applications

Organic–inorganic hybrid materials have attracted chemists, materials scientists, engineers, and physicists to exploit their potential applications in fields such as drug delivery, cosmetics, viscosity and surface modification, and water purification.

Amphiphilicity is an important character of many organic-inorganic hybrid materials, which may result in various self-assembly behaviors of these hybrid materials. The self-assembly of several POSS/polymer amphiphilic hybrid materials was systematically studied. POSS, or Polyhedral oligomer silsesquioxane is an organic inorganic hybrid material with a diameter of 1 nm as shown in Fig. 5, which make it a suitable building block for hybrid materials.

Hussain, H. et al.[39] synthesized amphiphilic poly(ethylene glycol)-b-poly-(pentafluorostyrene)-g-POSS (PEG-b-PPFS-g-POSS) hybrid copolymers, which were subsequently self-assembled in aqueous solutions to form micellar structures. The self-assembly behaviors of the hybrid materials were studied in depth by light scattering. In the work of B. H. Tan, et al.,[40,41] when hydrophobic POSS nanoparticles were added to block copolymer solutions of poly(ethylene glycol) (PEG) and poly(methacrylisobutyl polyhedral oligomericsilsesquioxane) P(MA-POSS) as the hydrophilic and hydrophobic blocks respectively, the micelle formation, gelation, and rheological performance of the amphiphilic copolymer could the tailored by the POSS nanoparticles. The amphiphilicity of the organic-inorganic hybrid materials could be helpful in medical applications. Fan, et al. synthesized a novel unimolecular star-like amphiphilic copolymer POSS-(PAA-(PLLA-PEG)$_4$)$_8$, which has eight arms of poly(acrylic acid)- (poly(L-lactide)-poly(ethylene glycol))$_4$ (PAA-(PLLA-PEG)$_4$).[42] Due to its unique architecture, this copolymer self-assembled into special nano-sized unimolecular micelles, which possess a PLA hydrophobic membrane

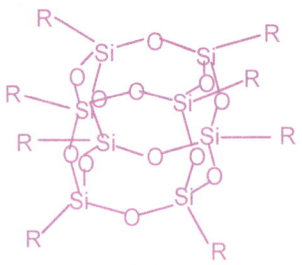

Fig. 5. Chemical Structure of POSS, R could be a reactive or non-reactive group.

with a PEG corona and interior PAA hydrophilic cavities. The PAA hydrophilic cavities can load some hydrophilic drugs via electrostatic interactions. Doxorubicin (DOX), a hydrophilic anticancer drug, was encapsulated into the micelles formed by POSS-(PAA-(PLLA-PEG)$_4$)$_8$ and the drug release profile was evaluated. In the work of Loh, et al.,[43] an amphiphilic star-like organic-inorganic cationic POSS-g-PDMAEMA polymer was synthesized and micelles could be formed by self-assembly of POSS-g-PDMAEMA. Paclitaxel, a tubulin-binding and anti-mitotic hydrophobic drug, was effectively loaded into the hydrophobic core (POSS) of the copolymer micelles. Amphiphilic organic-inorganic hybrid materials may also act as surfactants in particle transfer or separation applications. For example, in the synthesis of the Janus particle, it is always difficult to prepare such dipolar Janus nanoparticles in solution because both precursors are aqueous. Wang, et al.,[44] solved this problem by using commercial amphiphilic poly(ethylene glycol)-POSS (PEG-POSS). Assisted by PEG-POSS, negatively charged gold nanoparticles (Au NPs) were transferred into an organic solvent, retaining the Au NPs' negative charges in organic solvents.

pH or thermo responsiveness is another interesting character of organic-inorganic hybrid materials. Li, et al. synthesized a series of thermo-responsive amphiphilic hybrid copolymers with a random brush-like structure by copolymerizing hydrophilic poly(ethylene glycol)methacrylate (PEGMA) and hydrophobic POSS methacrylate (POSSMA) together with temperature sensitive poly- (propylene glycol)methacrylate (PPGMA).[45] The hybrid material can self-assemble to form micelles in aqueous media. The thermo-responsiveness of the PPG brush in the hybrid micelles was studied and it was found that the micelles could mimic the natural GroEL–GroES chaperone functionalities for renaturation of thermally denatured proteins in a simple and spontaneous "capture and release" method using easily controllable temperature as the sole trigger in an "on demand" fashion. In the work of Wang, et al.,[46] random-type amphiphilic pH-responsive hybrid copolymers with acrylic acid as pH-responsive hydrophilic and acrylate-POSS as hydrophobic constituents were studied. The hybrid copolymers exhibited pH-responsive self-assembly behavior in aqueous solutions. By molecular design, there could be pH and thermal dual-responsive organic-inorganic hybrid materials. Bai, et al., developed poly(acrylic acid-block-N-isopropylacrylamide) (PAAc-b-PNIPAm)-tethered POSS via POSS-initialized atom transfer radical polymerization.[47] The cloud point critical temperature (T_c) of POSS-PAAc-b-PNIPAm varies in a wide temperature range with pH. When the PAAc blocks are relatively long, the pH response of POSS-PAAc-b-PNIPAm diminishes owing to the reduced local PNIPAm chain density. Another pH and thermal dual-responsive hybrid material was also developed by Bai, et al.,[48] where the dendritic POSS-poly(N-isopropylacrylamide)

(PNIPAm)-poly(2-hydroxyethyl methacrylate) (PHEMA) copolymer demonstrated encapsulation of a fluorescent dye and stimulated release by temperature and pH changes.

Organic-inorganic hybrid materials also show potential as shape memory materials. In the work of Khine, et al.,[49,50] POSS-poly(ε-caprolactone) polyurethanes (POSS-PCL PU) with attractive shape memory properties was fabricated. The shape memory properties of POSS-PCL PU were studied by cyclic thermo-mechanical measurements. The experimental data unveiled enhanced shape fixities, recoverabilities and stress storage capacities at higher POSS content. The most promising POSS-PCL PU candidate exhibited remarkable cycle-averaged (N=2–5) shape fixities and strain recoverabilities of 98% and stress recoverabilities close to 100%.

Due to the interesting and even unique properties, organic-inorganic hybrid materials provide a wide range of possibilities in supramolecular templating applications. Supramolecular templating refers to a structure directing process making use of complex of molecules held together through non-covalent interactions. It basically involves two principal aspects, namely the self-assembly of organic templates that are composed of hydrophobic and hydrophilic portions, and the directing formation of inorganic or organic-inorganic hybrid nanostructures through favourable organic-inorganic interactions — either Coulomb interactions, hydrogen bonding force, or hydrophilic/hydrophobic interactions. Typical organic templates include small molecular surfactants such as cetyltrimethylammonium bromide (CTAB) and amphiphilic block copolymers such as polyethylene oxide-*block*-polypropylene oxide-*block*-polyethylene oxide (PEO-PPO-PEO). Their self-assembling aggregates such as micelles, micellar assemblies and liquid-crystal mesophases have been widely used to direct the formation of inorganic/hybrid nanostructures.

The NUS group is also working on hybrid nanostructured materials, in which a rich variety of titania and titania-based hybrid nanostructures have been designed and synthesized, including not only the simple morphologies (*e.g.*, spheres and vesicles), but also some complex superstructures (*e.g.*, mesoporous titania, large compound vesicles (LCVs) and hexagonally packed hollow hoops (HHHs)), through the supramolecular templating approach.[51–56] Studies have covered a wide range of processing conditions, synthesis mechanisms, characterization, and potential applications. The supramolecular-templated nanostructured titania are characteristic of large accessible internal surface area, narrow pore size distribution, regular pore organization with three-dimensional pore connectivity, an interconnected titania skeleton with regularly packed nanocrystalline anatase, and the capacity to incorporate ions and adsorbents. These features, combined with the semiconducting nature of titania, make them highly attractive for photocatalysis and dye-sensitized solar cell (DSSC) applications.[53,57–59] The pore organization and the crystallinity of titania were found to be key parameters in determining their photocatalytic activity

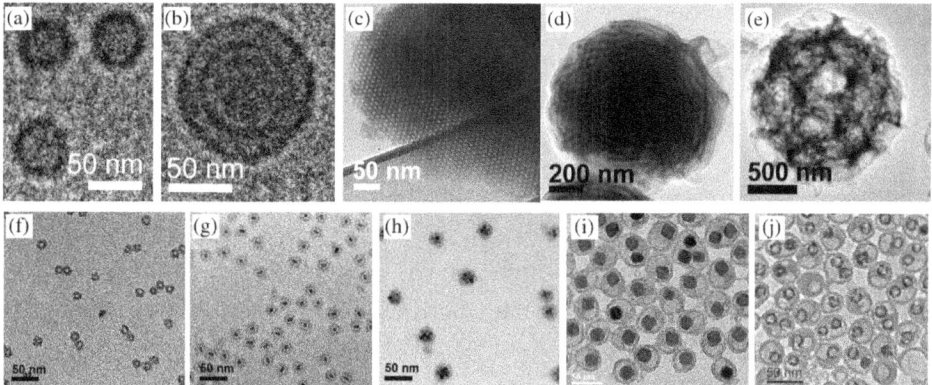

Fig. 6. Top: TEM micrographs of titania-based hybrid nanostructures: (a) sphere, (b) vesicle, (c) mesostructured titania, (d) LCVs and (e) HHHs. Bottom: TEM micrographs of silica-based hybrid nanostructures with encapsulation of (f) fluorescent conjugated polymers, (g) QDs, (h) Fe_3O_4 nanoparticles, (i) MnO and (j) hollow MnO nanocrystals.

and energy-conversion efficiency of solar cells. In addition, an interfacial templating condensation approach for the synthesis of polymer/silica hybrid nanocapsules (SNCs) has also been developed since 2009.[60] The basic idea of the synthesis is to deposit a silica shell through the sol-gel reactions at the core/corona interface of templating PEO-PPO-PEO micelles. As both silica and the templating polymers are US Food and Drug Administration (FDA)-approved materials, the hybrid SNCs have demonstrated to be excellent nanovehicles for encapsulation of a variety of bioactive agents including drugs,[60] fluorescent conjugated polymers,[60–63] quantum dots (QDs),[64] Fe_3O_4 nanoparticles, and MnO and hollow MnO nanocrystals for bioimaging/diagnosis and drug delivery/therapy applications.[61,65–68]

Functional Polymeric Nanocomposites

Electrically conductive epoxy-based hybrids filled with vapor grown carbon fibers (VGCF) and clay were developed by Kotaki, et al.[69] The VGCF-based hybrids with highly exfoliated clay have a low volume resistivity at relatively low VGCF contents, indicating a low percolation threshold. The low percolation threshold was achieved by an effective VGCF dispersion of electro-conductive network enhanced by highly exfoliated clay. In another study, a low-k polyimide (PI) nanocomposite from polyamic acid cured with POSS-OH was developed.[70] The incorporation of POSS-OH into a PI matrix reduced the dielectric constant of PI without losing the mechanical properties. In the work of Zhou, et al.,[71] POSS-OH, which has a large specific surface area and functionality number, was incorporated into electrospun poly(vinylidene fluoride-co-hexafluoropropylene) (PVDF-co-HFP) nanofibrous mats. The mats served to host $LiClO_4$/1-butyl-3-methylimidazolium tetrafluoroborate. It was found that POSS-OH can significantly increase both ionic conductivity and lithium transference number of the electrolytes owing to the Lewis acid-base interactions of POSS-OH with ClO_4^- and BF_4^-. Vadukumpully, et al. reported free standing ultrathin composite films of surfactant-wrapped graphene nanoflakes and poly(vinyl chloride) by a simple solution blending, drop casting and annealing route.[72] Mechanical properties and thermal stability of the polymer was enhanced by graphene. And the composite films had a very low percolation threshold of 0.6 vol% and showed a maximum electrical conductivity of 0.058 S/cm at 6.47 vol% of the graphene loading.

Today while we report our achievements on composites and hybrid materials in the past 25 years, our ultimate goal continues to be generating social and economic impact in addition to contributions to the scientific and engineering communities.

References and Notes

1. *Publication and citation information obtained via Thomson Reuters Web-of-Science online database search done on 25 August 2015.*
2. *http://www3.ntu.edu.sg/temasek-labs/advanced_material.html (Composites and advanced materials research at the Temasek Laboratory in Nanyang Technological University, TL@NTU).*
3. Loo, S.-L., et al., Superabsorbent cryogels decorated with silver nanoparticles as a novel water technology for point-of-use disinfection. Environmental science & technology, 2013. **47**(16): pp. 9363–9371.
4. *Spongy Material Puts The Squeeze On Waterborne Disease: Water Purification — A porous gel decorated with silver nanoparticles delivers disinfected water in seconds, (http://cen.acs.org/articles/91/web/2013/08/Spongy-Material-Puts-Squeeze-Waterborne.html) Chemical and Engineering News (C&EN), American Chemical Society (ACS), (August 21, 2013), Editorial news piece by Deirdre Lockwood.*
5. *Dirty water gets squeezed clean, Nature, 502, 145, (10 October 2013), doi:10.1038/502145f*
6. *Loo, S.-L., et al., Bactericidal Mechanisms Revealed for Rapid Water Disinfection by Superabsorbent Cryogels Decorated with Silver Nanoparticles.* Environmental science & technology, 2015. **49**(4): pp. 2310–2318.
7. Hussain, F., et al., Review article: polymer-matrix nanocomposites, processing, manufacturing, and application: an overview. Journal of composite materials, 2006. **40**(17): pp. 1511–1575.
8. Huang, S., et al., Morphology, thermal, and rheological behavior of nylon 11/multiwalled carbon nanotube nanocomposites prepared by melt compounding. Polymer Engineering and Science, 2009. **49**(6): p. 1063.

9. Liu, T., et al., *Morphology and mechanical properties of multiwalled carbon nanotubes reinforced nylon-6 composites.* Macromolecules, 2004. **37**(19): pp. 7214–7222.
10. Zhang, W.D., et al., *Carbon nanotubes reinforced nylon-6 composite prepared by simple melt-compounding.* Macromolecules, 2004. **37**(2): pp. 256–259.
11. Jin, Z., et al., *Dynamic mechanical behavior of melt-processed multi-walled carbon nanotube/poly (methyl methacrylate) composites.* Chemical Physics Letters, 2001. **337**(1): pp. 43–47.
12. Jin, Z., et al., *Poly (vinylidene fluoride)-assisted melt-blending of multi-walled carbon nanotube/poly (methyl methacrylate) composites.* Materials Research Bulletin, 2002. **37**(2): pp. 271–278.
13. Wang, S.-F., et al., *Preparation and mechanical properties of chitosan/carbon nanotubes composites.* Biomacromolecules, 2005. **6**(6): pp. 3067–3072.
14. Shen, L., et al., *Nanoindentation and morphological studies of epoxy nanocomposites.* Macromolecular Materials and Engineering, 2006. **291**(11): pp. 1358–1366.
15. Wang, S., et al., *Biopolymer chitosan/montmorillonite nanocomposites: preparation and characterization.* Polymer Degradation and Stability, 2005. **90**(1): pp. 123–131.
16. He, C., et al., *Microdeformation and fracture mechanisms in polyamide-6/organoclay nanocomposites.* Macromolecules, 2008. **41**(1): pp. 193–202.
17. Liu, T., et al., *A processing-induced clay dispersion and its effect on the structure and properties of polyamide 6.* Polymer International, 2004. **53**(4): pp. 392–399.
18. Pramoda, K., et al., *Thermal degradation behavior of polyamide 6/clay nanocomposites.* Polymer degradation and Stability, 2003. **81**(1): pp. 47–56.
19. Phang, I.Y., et al., *Morphology, thermal and mechanical properties of nylon 12/organoclay nanocomposites prepared by melt compounding.* Polymer international, 2005. **54**(2): pp. 456–464.
20. Wang, L., et al., *Hydrothermal effects on the thermomechanical properties of high performance epoxy/clay nanocomposites.* Polymer Engineering & Science, 2006. **46**(2): pp. 215–221.
21. Phonthammachai, N., et al., *Fabrication of CFRP from high performance clay/epoxy nanocomposite: Preparation conditions, thermal–mechanical properties and interlaminar fracture characteristics.* Composites Part A: Applied Science and Manufacturing, 2011. **42**(8): pp. 881–887.
22. Wang, K., et al., *Preparation, microstructure and thermal mechanical properties of epoxy/crude clay nanocomposites.* Composites Part A: applied science and manufacturing, 2007. **38**(1): pp. 192–197.
23. Wang, L., et al., *Preparation, morphology and thermal/mechanical properties of epoxy/nanoclay composite.* Composites Part A: Applied Science and Manufacturing, 2006. **37**(11): pp. 1890–1896.
24. Wang, K., et al., *Epoxy nanocomposites with highly exfoliated clay: mechanical properties and fracture mechanisms.* Macromolecules, 2005. **38**(3): pp. 788–800.
25. Mya, K.Y., et al., *The effect of nanofiller on the thermomechanical properties of polyimide/clay nanocomposites.* Macromolecular Chemistry and Physics, 2008. **209**(6): pp. 643–650.

26. Bao, Q., et al., *Graphene–polymer nanofiber membrane for ultrafast photonics.* Advanced Functional Materials, 2010. **20**(5): pp. 782–791.
27. Sin, S.L., et al., *The Effect of Different Clay Dispersion Methods on the Properties of Polyurethane/Clay Nanocomposites.* Australian Journal of Chemistry, 2013. **66**(9): pp. 1039–1047.
28. Pramoda, K., et al., *Covalent bonded polymer–graphene nanocomposites.* Journal of Polymer Science Part A: Polymer Chemistry, 2010. **48**(19): pp. 4262–4267.
29. Huang, J., et al., *Organic–inorganic nanocomposites from cubic silsesquioxane epoxides: direct characterization of interphase, and thermomechanical properties.* Polymer, 2005. **46**(18): pp. 7018–7027.
30. Huang, J.-c., et al., *Polyimide/POSS nanocomposites: interfacial interaction, thermal properties and mechanical properties.* Polymer, 2003. **44**(16): pp. 4491–4499.
31. Yang, B.X., et al., *Mechanical Reinforcement of Polyethylene Using Polyethylene-Grafted Multiwalled Carbon Nanotubes.* Advanced Functional Materials, 2007. **17**(13): pp. 2062–2069.
32. Sun, Y. and C. He, *Biodegradable "core–shell" rubber nanoparticles and their toughening of poly (lactides).* Macromolecules, 2013. **46**(24): pp. 9625–9633.
33. Sun, Y., et al., *Biodegradable and renewable poly (lactide)–lignin composites: synthesis, interface and toughening mechanism.* Journal of Materials Chemistry A, 2015. **3**(7): pp. 3699–3709.
34. Thitsartarn, W., et al., *Simultaneous Enhancement of Strength and Toughness of Epoxy using POSS-Rubber Core-Shell Nanoparticles.* Composites Science and Technology, 2015.
35. *Patent filed: "Novel Fillers for Strong and Tough Polymer Composite" 2015.*
36. Fujihara, K., M. Kotaki, and S. Ramakrishna, *Guided bone regeneration membrane made of polycaprolactone/calcium carbonate composite nano-fibers.* Biomaterials, 2005. **26**(19): pp. 4139–4147.
37. Prabhakaran, M.P., J. Venugopal, and S. Ramakrishna, *Electrospun nanostructured scaffolds for bone tissue engineering.* Acta Biomaterialia, 2009. **5**(8): pp. 2884–2893.
38. Gupta, D., et al., *Nanostructured biocomposite substrates by electrospinning and electrospraying for the mineralization of osteoblasts.* Biomaterials, 2009. **30**(11): pp. 2085–2094.
39. Hussain, H., et al., *Synthesis, micelle formation, and bulk properties of poly (ethylene glycol)-b-poly (pentafluorostyrene)-g-polyhedral oligomeric silsesquioxane amphiphilic hybrid copolymers.* Journal of Polymer Science Part A: Polymer Chemistry, 2010. **48**(1): pp. 152–163.
40. Hussain, H., et al., *Micelle formation and gelation of (PEG– P (MA-POSS)) amphiphilic block copolymers via associative hydrophobic effects.* Langmuir, 2010. **26**(14): pp. 11763–11773.
41. Tan, B., H. Hussain, and C. He, *Tailoring Micelle Formation and Gelation in (PEG–P (MA-POSS)) Amphiphilic Hybrid Block Copolymers.* Macromolecules, 2011. **44**(3): pp. 622–631.

42. Fan, X., Z. Wang, and C. He, "Breathing" unimolecular micelles based on a novel star-like amphiphilic hybrid copolymer. Journal of Materials Chemistry B, 2015.
43. Loh, X.J., et al., Efficient gene delivery with paclitaxel-loaded DNA-hybrid polyplexes based on cationic polyhedral oligomeric silsesquioxanes. Journal of Materials Chemistry, 2010. **20**(47): pp. 10634–10642.
44. Wang, F., et al., PEG-POSS Assisted facile preparation of amphiphilic gold nanoparticles and interface formation of Janus nanoparticles. Chemical Communications, 2011. **47**(2): pp. 767–769.
45. Li, Z., et al., Design of polyhedral oligomeric silsesquioxane (POSS) based thermo-responsive amphiphilic hybrid copolymers for thermally denatured protein protection applications. Polymer Chemistry, 2014. **5**(23): pp. 6740–6753.
46. Wang, Z., et al., pH-responsive amphiphilic hybrid random-type copolymers of poly (acrylic acid) and poly (acrylate-POSS): synthesis by ATRP and self-assembly in aqueous solution. Colloid and Polymer Science, 2013. **291**(8): pp. 1803–1815.
47. Bai, Y., et al., Temperature and pH dual-responsive behavior of polyhedral oligomeric silsesquioxane-based star-block copolymer with poly (acrylic acid-block-N-isopropylacrylamide) as arms. Colloid and Polymer Science, 2012. **290**(6): pp. 507–515.
48. Bai, Y., et al., Temperature and pH Dual-Responsive Behavior of Dendritic Poly (N-isopropylacrylamide) with a Polyoligomeric Silsesquioxane Core and Poly (2-hydroxyethyl methacrylate) Shell. Macromolecular Chemistry and Physics, 2013. **214**(3): pp. 396–404.
49. Bothe, M., et al., Triple-shape properties of star-shaped POSS-polycaprolactone polyurethane networks. Soft Matter, 2012. **8**(4): pp. 965–972.
50. Mya, K.Y., et al., Star-shaped POSS-polycaprolactone polyurethanes and their shape memory performance. Journal of Materials Chemistry, 2011. **21**(13): pp. 4827–4836.
51. Yuwono, A.H., et al., Diblock copolymer templated nanohybrid thin films of highly ordered TiO2 nanoparticle arrays in PMMA matrix. Chemistry of materials, 2006. **18**(25): pp. 5876–5889.
52. Zhang, Y., et al., Sonochemical synthesis and liquid crystal assembly of PS-b-PEO–titania aggregates. Chemical Communications, 2012. **48**(68): pp. 8538–8540.
53. Zhang, Y., J. Li, and J. Wang, Substrate-assisted crystallization and photocatalytic properties of mesoporous TiO2 thin films. Chemistry of materials, 2006. **18**(12): pp. 2917–2923.
54. Zhang, Y., et al., Highly dispersed gold nanoparticles assembled in mesoporous titania films of cubic configuration. Microporous and Mesoporous Materials, 2008. **110**(2): pp. 242–249.
55. Zhang, Y., et al., Formation and Evolution of Body-Centered Orthorhombic Mesophase in TiO2 Thin Films. Journal of the American Ceramic Society, 2009. **92**(6): pp. 1317–1321.
56. Zhang, Y., et al., Hybrid titania microspheres of novel superstructures templated by block copolymers. Chemistry of Materials, 2011. **23**(11): pp. 2745–2752.
57. Zhang, Y., et al., Enhanced photocatalysis by doping cerium into mesoporous titania thin films. The Journal of Physical Chemistry C, 2009. **113**(51): pp. 21406–21412.

58. Zhang, Y., Z. Xie, and J. Wang, *Supramolecular-templated thick mesoporous titania films for dye-sensitized solar cells: Effect of morphology on performance.* ACS applied materials & interfaces, 2009. **1**(12): pp. 2789–2795.
59. Zhang, Y., Z. Xie, and J. Wang, *Highly efficient dye-sensitized solar cells of thick mesoporous titania films derived from supramolecular templating.* Nanotechnology, 2009. **20**(50): pp. 505602.
60. Tan, H., et al., *Facile synthesis of hybrid silica nanocapsules by interfacial templating condensation and their application in fluorescence imaging.* Chemical Communications, 2009(41): pp. 6240–6242.
61. Tan, H., et al., *Silica Nanocapsules of Fluorescent Conjugated Polymers and Superparamagnetic Nanocrystals for Dual-Mode Cellular Imaging.* Chemistry-A European Journal, 2011. **17**(24): pp. 6696–6706.
62. Tan, H., et al., *Silica-shell cross-linked micelles encapsulating fluorescent conjugated polymers for targeted cellular imaging.* Biomaterials, 2012. **33**(1): pp. 237–246.
63. Hsu, B.Y.W., et al., *PEO surface-decorated silica nanocapsules and their application in in vivo imaging of zebrafish.* RSC Advances, 2012. **2**(32): pp. 12392–12399.
64. Zhang, Y., et al., *PEOlated micelle/silica as dual-layer protection of quantum dots for stable and targeted bioimaging.* Chemistry of Materials, 2013. **25**(15): pp. 2976–2985.
65. Tan, H., et al., *Synthesis of PEOlated Fe3O4@ SiO2 nanoparticles via bioinspired silification for magnetic resonance imaging.* Advanced Functional Materials, 2010. **20**(5): pp. 722–731.
66. Hsu, B.Y.W., et al., *Silica–F127 nanohybrid-encapsulated manganese oxide nanoparticles for optimized T 1 magnetic resonance relaxivity.* Nanoscale, 2014. **6**(1): pp. 293–299.
67. Zhang, Y., et al., *Silica-based nanocapsules: synthesis, structure control and biomedical applications.* Chemical Society Reviews, 2015. **44**(1): pp. 315–335.
68. Hsu, B.Y.W., et al., *A Hybrid Silica Nanoreactor Framework for Encapsulation of Hollow Manganese Oxide Nanoparticles of Superior T1 Magnetic Resonance Relaxivity.* Advanced Functional Materials, 2015. **25**(33): pp. 5269–5276.
69. Kotaki, M., et al., *Electrically conductive epoxy/clay/vapor grown carbon fiber hybrids.* Macromolecules, 2006. **39**(3): pp. 908–911.
70. Wahab, M.A., K.Y. Mya, and C. He, *Synthesis, morphology, and properties of hydroxyl terminated-POSS/polyimide low-k nanocomposite films.* Journal of Polymer Science Part A: Polymer Chemistry, 2008. **46**(17): pp. 5887–5896.
71. Zhou, R., et al., *Electrospun poly (vinylidene fluoride) copolymer/octahydroxypolyhedral oligomeric silsesquioxane nanofibrous mats as ionic liquid host: enhanced salt dissociation and its function in electrochromic device.* Electrochimica Acta, 2014. **146**: pp. 224–230.
72. Vadukumpully, S., et al., *Flexible conductive graphene/poly (vinyl chloride) composite thin films with high mechanical strength and thermal stability.* Carbon, 2011. **49**(1): pp. 198–205.

3. Materials for Water Remediation (Membranes)

Sui Zhang*, Lin Luo, Zhi Wei Thong and Tai-Shung Chung†

1. Introduction

As a highly-populated city of five million people, Singapore's current demand for over 1.7 million cubic meters of water per day is expected to double in the next 50 years. The land area is approximately 700 km² in Singapore. Although rainfall reaches 250 cm per year over the island, water is a scarce resource due to the limited space for water collection and storage. (PUB, 2011). Other factors that hamper the availability of water include the fact that Singapore lacks natural aquifers and groundwater, and yet is surrounded by the sea.

Being aware of the strategic importance of water, Singapore strived to build a robust water supply through planning and investment in research and technologies over the last 50 years. The water supply is now diversified across four different sources, including local catchment water; imported water; reclaimed water known as NEWater which meets the drinking water standard; and desalinated water. A schematic of the water system is shown in Fig. 1 (PUB, 2011; Wachinski, 2013). Currently there are 17 reservoirs over the island for water storage, which occupy two thirds of the total land area. They play an important role in Singapore's water cycle.

NEWater and desalinated water supplement the water supply from unconventional sources, such as wastewater and seawater. The efforts in fact dated back to

---

* chezhangsui@nus.edu.sg.
† chencts@nus.edu.sg.
Authors are from the Department of Chemical and Biomolecular Engineering, National University of Singapore.

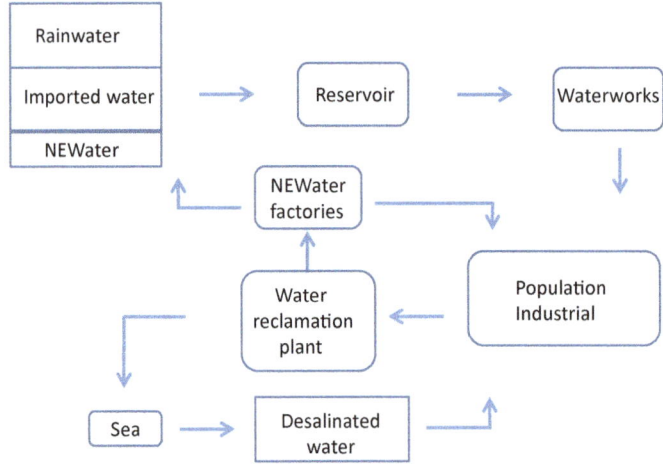

Fig. 1. A schematic of Singapore's water system.

the 1970s. In 2003, the first NEWater plants were opened in Bedok and Kranji to get clean water from municipal and industrial wastewaters. The number of such plants was increased to 5 in 2010. The treatment processes include pretreatment by activated sludge, micro-screening, membrane separation processes (microfiltration, MF; or ultrafiltration, UF; and reverse osmosis, RO), and UV disinfection. The reclaimed water is of a high quality with the turbidity less than 0.5 ntu, the total organic carbon less than 0.5 mg L^{-1}, and the TDS less than 50 mg L^{-1} (Wachinski, 2013), which meets the drinking water standard adopted by the World Health Organization. Most of the product water is supplied to local industries for non-potable usage, and some is discharged into the reservoirs for possible use in the dry periods. Now the combined capacity of the five plants reaches to 459,000 m^3 d^{-1}, providing 30% of Singapore's water needs. By 2060, the NEWater capacities are projected to meet up to 55% of the nation's future water consumption. (PUB, 2011)

The first desalination plant SingSpring was started slightly later in 2005 in Tuas. As one of Asia's largest seawater desalination RO plants, it produces 136,000 cubic meters of water per day. A second one was followed in 2013. It is projected that in 2060 desalinated water should meet 25% of the total water demand. (PUB, 2013)

Both approaches rely on the modern membrane technologies, albeit traditional MF, UF and RO processes have been involved. To improve the efficiency of the current systems, and also to build Singapore as the global hydrohub for leading-edge technologies, both the Public Utilities Board (PUB) and the National Research Foundation (NRF) have continuously invested on membrane research and development. Under the leadership of the PUB chairman, Mr. Tan Gee Paw, and the Chief Technology Officer, Mr. Harry Seah, research and applications have been very active towards the design of novel membranes and membrane processes for water treatment in Singapore, especially in the last 10 years. This chapter aims to summarize Singapore's major achievements on various developments in membrane technology.

2. Overview of Membranes and Membrane Processes for Water Remediation

In a typical membrane separation process, an effective driving force such as a concentration, a temperature or a pressure difference, is applied upon the feed mixture to achieve preferential transport of one or more components across the membrane. Compared to conventional processes, membrane separation is more compact, energy efficient and cost effective. It can have wide applications in the areas of energy, water production, pharmaceutical and life sciences. The modern membrane industry began in the 1960s, when Loeb and Sourirajan invented the phase inversion method to manufacture asymmetric cellulose acetate membranes for seawater desalination. By the 1980s, MF, UF and RO processes were all established. From the late 1990s, treatment of the municipal wastewater by MF/UF became commercially available (Baker, 2012). The last ten years witnessed the explosive progress in the research of membrane distillation (MD) and forward osmosis (FO) processes. The key challenge for the development of these processes is the membrane. Today, not only has Singapore perfected MF and UF membranes, but the country also has the most advanced technologies for MD and FO membranes.

Membranes can be categorized according to their materials, types (structures) or processes. A general introduction to the membranes for water remediation and Singapore's contributions to the field will be given in the following sections.

2.1 Membrane materials

The majority of commercially available membranes are made from organic polymers. For most applications except membrane distillation, relatively hydrophilic polymers are preferred. Blending, modifications or surface coating by another type of highly hydrophilic molecules upon the base membranes have been practiced to enhance the anti-fouling properties of the membranes. The molecules can be polymers or inorganic nanoparticles. In recent years, ceramic membranes are used in some UF or MF processes where solvent resistance and thermal stability are required.

2.2 Membrane types

Depending on the molecular homogeneity, membranes can be classified as isotropic or anisotropic ones. Fig. 2a illustrates the schematics of their structures. Dense isotropic films are seldom used for water purification due to the high resistance of water transport across the thick films. A microporous isotropic membrane is similar to the conventional filter in terms of structure and function, but with a much smaller pore size of less than 10 µm in diameter. Particles larger than the membrane pore size are rejected by the size exclusion mechanism.

To reduce the transport resistance from the barrier membrane but maintaining reasonable mechanical stability, anisotropic membranes have been developed by various fabrication methods. Usually, they consist of a thin selective layer and a thick porous support. The two layers can be formed in a single or multi-step operation. Temperature- or nonsolvent-induced phase inversion methods are typical ways to produce such structures in a single operation. Composite membranes are formed by separate preparation of the support and then the selective layer, which allows the independent control of the two layers.

On the other hand, membranes can also be classified as flat sheets or hollow fibers depending on the configuration. Fig. 2b provides the SEM images of typical flat sheet and hollow fiber membranes.

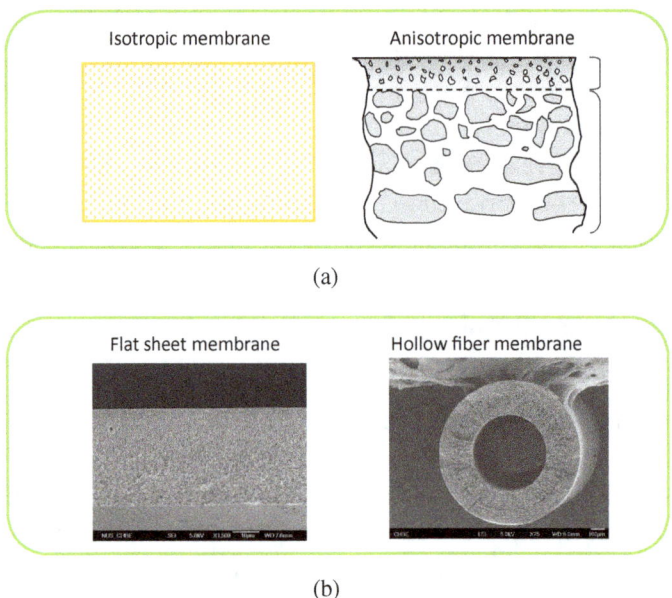

Fig. 2. Schematics of (a) isotropic and anisotropic membranes and (b) flat sheet and hollow fiber membranes.

2.3 Membrane processes

Based on the pore size of the selective layer, traditional pressure-driven membrane processes can be divided into MF, UF, nanofiltration (NF) or RO. Fig. 3 gives a schematic illustration. MF removes large particles such as bacteria from the feed and UF is able to separate most viruses from water. Both processes usually serve as the pretreatment steps for RO. NF membranes are used to remove dyes, pharmaceuticals and divalent ions, and RO is applied for desalination. In addition, MF or UF processes can be integrated with bioreactors to form membrane bioreactors (MBR).

Both MD and FO are emerging desalination technologies in recent decades. MD is driven by the difference in water vapor pressure across the membrane at different temperatures. It has a pore size similar to MF ones. FO utilizes the osmotic pressure gradient between different salinity solutions for desalination, wastewater treatment or power generation. Depending on the applications, the pore size of FO membranes may vary from dense RO types to loose NF types.

2.4 Applications and research in Singapore

MF, UF and RO have been widely employed in Singapore for water production by the 5 NEWater plants and 2 desalination plants over the island. Since 2003, three membrane bioreactor (MBR) pilot plants have been run by PUB. Encouraged by the high quality output, Singapore is set to introduce MBR as a key technology to improve the efficiency and reduce the cost for water reclamation (PUB, 2011). In addition, local enterprises are actively participating in R & D of various membrane processes for water remediation. A MD pilot plant was built in 2010 (PUB, 2010).

The rapid commercialization of membrane processes in Singapore has supported and is in return boosted by intense research on novel materials and membranes in

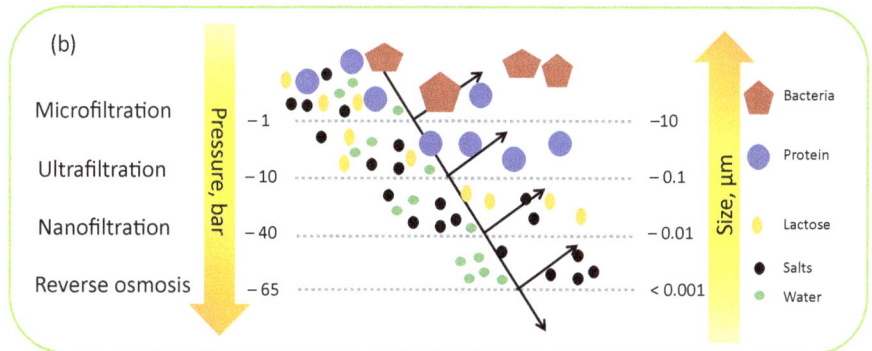

Fig. 3. An illustration of the pressure-driven membrane separation processes.

local universities and institutions. This review has a wide coverage of the technology available, ranging from low pressure-driven MF/UF/MBR and high pressure-driven NF/RO membranes, to the newly emerging MD and FO membranes.

3. Membranes for Low Pressure-Driven Processes

3.1 The low pressure-driven processes

MF and UF are typical low pressure-driven membrane separation processes. The transmembranous pressure works as the driving force for fluid transport, which forces the fluid and small solutes to pass through and be collected as the permeate; whereas large molecules are rejected and retained as the retentate. Typical operating pressures range from 0.2 to 6 bar. Filtration membranes are characterized by pore sizes in the range of 1~100 nm for UF, and 0.1~10 μm for MF.

The basic separation mechanism of MF and UF membranes is molecular sieving (size exclusion). Flux and rejection are two most important performance parameters. Flux is defined as the permeated volume across the membrane per unit time per unit membrane area. A higher flux usually brings lower capital and operating costs. Rejection is a measure of a membrane's separation capabilities, which can be expressed as the nominal molecular weight cut-off (MWCO). MWCO is defined as the molecular weight of a test solute which is 90% rejected by the membrane under standard conditions.

The membrane filtration equipment for water and wastewater treatment in Southeast Asia had a business size of $249.1 million in 2011 and is expected to reach $398.0 million by 2017. In 2011, MF/UF occupied 60% of the total market size and Singapore made up almost one quarter of the Southeast Asian market (Frost & Sullivan Market Insight, 2011). MF/UF membrane technologies have played vital roles for wastewater reclamation in Singapore. They are also used as the pre-treatment processes for RO plants to produce NEWater and desalinated water (PUB, 2011).

3.2 Introduction to MF/UF membranes

MF/UF membranes can be made from either polymeric or inorganic materials. Polymers such as cellulose acetate (CA), polyamide (PA), polysulfone (PSf), polyethersulfone (PES), polyvinylidene fluoride (PVDF), and polypropylene (PP) have been utilized. They are relatively of low cost and are easy to manufacture. However, they are sensitive to the pH, temperature and chlorine in the feed solution. Inorganic membranes, including alumina, glass and zirconia, find usage in corrosive applications with their drawbacks being their high manufacturing costs and brittleness (Baker, 2012). Globally, about 80 to 90% of the installed membranes are polymeric.

In Singapore, commercial MF/UF membranes available in the local market include Memcor's CMF-S system, GE's ZeeWeed UF membranes, Hyflux's Kristal® hollow fiber membranes, etc. Research on new MF/UF membranes mainly focuses on high flux and low fouling ones. Fouling not only increases the frequency of membrane cleaning and energy costs but may also require frequent membrane replacement to maintain the productivity. To reduce fouling, current MF/UF membranes are relatively hydrophilic with contact angles of 60~80°. However, contact angles of less than 30° are more desirable (Fane et al., 2015).

3.3 High-flux MF/UF membranes

A higher permeate flux or a better solute rejection can help save energy and reduce the capital cost. Useful strategies to increase water flux involve alterations in porosity, pore size, membrane thickness and hydrophilicity. For example, the MF membrane fabricated from electro-spun PVDF nanofibers exhibited high pure water permeability of more than 10000 L h^{-1} m^2 bar^{-1} (Gopal et al., 2006) due to their large pore size (4~10.6 μm) and high porosity. In addition, blending PVDF with surface modifying macromolecules may increase the hydrophilicity of the nanofibers (Kaur et al., 2012). However, weak mechanical strength and the possible reduction in porosity in the long run are main concerns of the electro-spun nanofibers.

Hollow fiber membranes are of great interest for practical applications because of their self-support nature, easy module fabrication and high packing density. Early studies paid great attention to the spinning parameters that lead to high performance hollow fiber membranes for UF. Fig. 4 shows some key spinning parameters

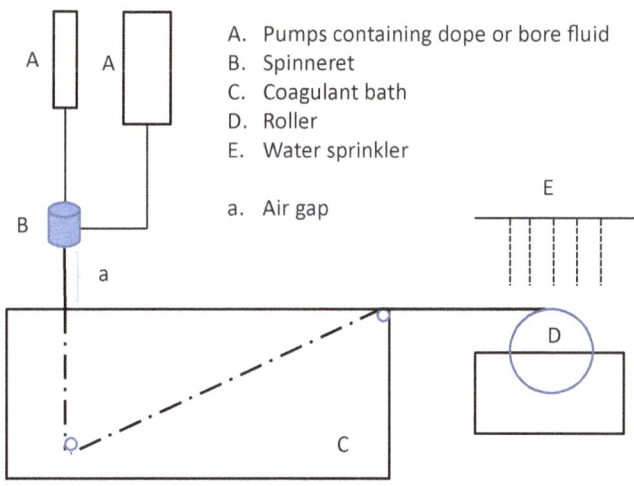

Fig. 4. Schematic diagram of a hollow fiber spinning line.

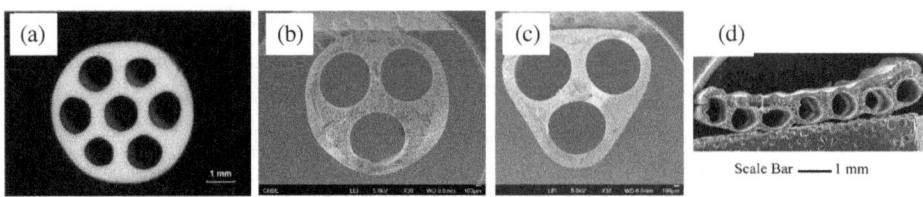

Fig. 5. Cross-section images of the (a) 7-bore (Wang and Chung, 2015), (b) round tribore (Luo et al., 2014), (c) triangular tribore (Wang et al., 2014) and (d) rectangular (Peng et al., 2011) ultrafiltration hollow fiber membranes.

and NUS facilities for hollow fiber spinning. Singapore scientists have found that a higher dope flow rate (i.e., shear rate) in the spinneret tended to result in membranes with a smaller mean pore size and a denser skin due to the shear-induced orientation (Qin and Chung, 1999). Air gap greatly influenced hollow fiber formation and morphology (Qin et al., 2001). The flow angle in spinnerets also played an important role in determining the pore size of membranes (Wang et al., 2004).

Recently, multi-bore hollow fiber membranes (Fig. 5) have emerged with superior mechanical properties, higher packing density and enhanced long-term durability. Inge GmbH launched the Multibore® PES hollow fiber membrane for UF (inge GmbH, 2013). With seven inner bores and a round-shape configuration, this membrane exhibits improved operation stability and reduced fiber breakage. Based on the pilot study in the Addur SWRO desalination plant, Bahrain, the Multibore® membranes outperformed conventional spiral-wound UF membranes, with lower fouling tendency, and less chemical and energy consumption (Bu-Rashid and Czolkoss, 2007). Meanwhile, the tri-bore hollow fiber (TBF) PVDF or PES membranes, named as Kristal®, were commercialized by Hyflux and applied for seawater pretreatment in both the world's largest membrane-based seawater desalination plant (in Magtaa, Algeria) and China's largest such plant in Tianjin. Kristal® UF membranes were believed to have an enhanced flux, improved system recovery and low fouling tendency (Hyflux, 2008). However, the main drawback of this round geometry is the non-uniform wall thickness, which causes additional resistance for water transport across the thicker portion along the fiber wall. To solve this problem, triangle-shaped membranes with a much more uniform wall thickness were developed by NUS (Luo et al., 2014, Wang et al., 2014b). The triangular TBF membranes exhibited higher water fluxes than round ones while at the same time maintained excellent mechanical strength. Due to the reduced cross sectional area, a higher packing density was theoretically allowed (Luo et al., 2014). Besides, rectangular seven-bore membranes based on PAN were developed by NUS which possessed the hybrid properties of flat sheet membranes and hollow fibers (Peng et al., 2011).

3.4 Low-fouling MF/UF membranes

Fouling in filtration systems is affected by a variety of parameters, such as permeate flux, operation mode, pressure, feed quality, membrane properties (hydrophilicity, roughness, pore size, charge), etc. Increasing membrane hydrophilicity is a commonly adopted method to reduce fouling. Theoretically, hydrophilic membranes tend to bind a thin layer of water molecules on the surface, which hinders the adhesion of hydrophobic foulants. As summarized in Table 1, strategies such as coating, blending, and chemical modifications have been utilized to enhance the hydrophilicity of MF/UF membranes (Liu, 2014).

Using hydrophilic materials to manufacture UF membranes is the most direct way. For example, sulfonated polyphenylenesulfone (sPPSU) polymers synthesized by BASF via the directly copolymerized sulfonation method were used to fabricate tri-bore hollow fibers by NUS and BASF scientists. The fibers exhibited lower fouling tendency and higher permeate flux than PPSU ones for oil/water emulsion separation. The analysis by the resistance-in-series model revealed that the resistances from surface adsorption and irreversible fouling were minimized in the sPPSU membrane due to the increased hydrophilicity (Luo et al., 2015).

An alternative method is to blend hydrophilic polymers or inorganic matters into the polymeric membrane matrix. By incorporating amphiphilic Pluronic block copolymers into PVDF, NTU scientists found that the resultant membranes became more porous and hydrophilic with a larger water flux (Loh and Wang, 2014). A hydrophilic and oleophobic PVDF membrane was also developed, in NUS, by blending PVDF with additives (Zhu et al., 2013). Graphene oxide (GO)-imbedded nanocomposite sPPSU hollow fiber membranes was another example that showed substantial reductions in reversible, irreversible and adsorption-induced fouling in oil/water separations (Tang et al., 2015).

Table 1. A Brief Summary of the Methods for Surface Modifications of Microfiltration/Ultrafiltration Membranes.

Method	Description	Modifying Materials (examples)
Blending/Mixed matrix	Two or more materials are physically mixed to form a layer	Polyvinylpyrrolidone, GO, etc.
Coating	A thin layer noncovalently adheres to the surface. Dip coating, spin coating and other approaches may be applied.	Chitosan, GO-TiO_2 microspheres, etc.
Chemical	Functional molecules are covalently introduced to the surface.	Polydopamine, polyelectrolytes, etc.

Additionally, Singapore scientists have found that physically coating a thin layer onto the membrane surface helps tailor the surface hydrophilicity, charge and roughness. A chitosan polymer was coated on the surface of PVDF membranes (Nasreen et al., 2014). Similarly, inorganic materials were also used. Silver nanoparticle/multiwalled carbon nanotubes (Ag/MWNTs) were coated as a disinfection layer to effectively inhibit the growth of bacteria in the filtration module and to prevent the formation of biofilm on the membrane surface (Gunawan et al., 2011). In addition, TiO_2 was widely adopted for its hydrophilicity and photocatalytic degradation of bio-pollutants. A novel GO–TiO_2 microsphere hierarchical membrane was fabricated by assembling the GO–TiO_2 composite microspheres on the surface of cellulose acetate membranes. The membranes showed high permeate flux, high photodegradation activity and negligible fouling (Gao et al., 2013).

Chemical modifications may provide membranes with better long-term stability and certain reaction-permissible surfaces. Poly(dopamine acrylamide)-grafted PVDF (PVDF-g-PDA) MF membranes have served as a versatile platform for subsequent modifications (Xu et al., 2013). An electrophoresis-UV grafting method was applied to bind polyelectrolytes onto PES membranes for the surface water treatment of a Singapore reservoir (Wei et al., 2006). Lower natural organic matter fouling was observed due to their hydrophilic and negatively charged characteristics. In addition, zwitterionic poly(sulfobetaine methacrylate) [poly(SBMA)] was tethered onto a polypropylene membrane surface and the resultant PP membrane had a water contact angle of 17.4° with reduced fouling (Zhao et al., 2010).

4. Membranes for High Pressure-Driven Processes

4.1 Introduction

RO is a high pressure-driven process that employs a membrane permeable to water but impermeable to salts. As a mature technology, RO has gained worldwide acceptance for brackish water and seawater desalination. All NEWater and desalination plants in Singapore adopt the RO systems to produce high quality water. The RO membranes have undergone rapid development since the early 1960s, when asymmetric cellulose acetate (CA) membranes were made by Loeb and Sourirajan via the phase inversion method (Loeb and Sourirajan, 1964). Later, aromatic thin film composite (TFC) membranes with good salt rejection and high water flux were invented by Cadotte (1981), which became the new industry standard up to now. RO research in Singapore started on the study of fouling phenomena and gradually focused on biomimetic and other membranes (Wang H et al., 2012; Wang H et al., 2013; Fane et al., 2015).

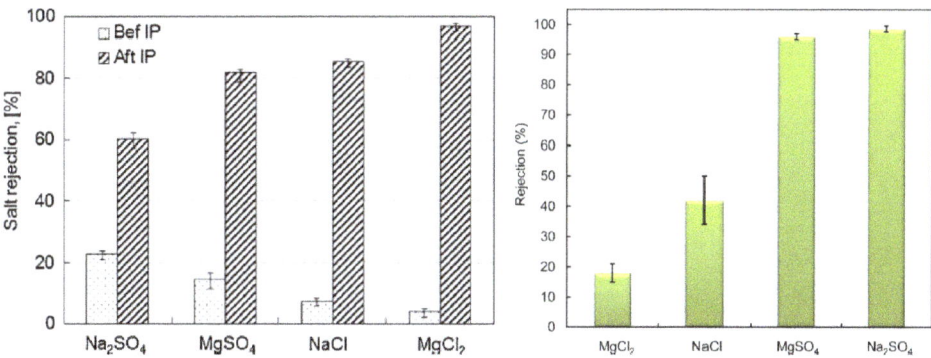

Fig. 6. The rejection to typical salts in aqueous solutions by (left) a positively charged membrane (Sun et al., 2012) and (right) a negatively charged membrane (Ong and Chung, 2014b).

Another high pressure-driven process is NF. NF membranes have a nominal molecular weight cutoff ranging from 200 to 1000Da and pore size ranging from 0.5 to 2.0nm. The separation mechanisms include size exclusion, Donnan effects (charge repulsion) and solute-membrane affinity. A comparison of rejections to typical ions in water by membranes with different charge characteristics is provided by NUS scientists as shown in Fig. 6. The left one refers to the positively charge membrane which shows a high rejection to Mg^{2+} (Sun et al., 2012), and the right one refers to the negatively charged membrane which effectively rejects SO_4^{2-} (Ong and Chung, 2014b).

The application range of NF has been growing rapidly since its introduction in the late 1980s. It was initially employed in water softening and organics removal (Eriksson, 1988). For example, NF membranes were used to soften municipal water by removing divalent ions or as a pre-treatment step for an RO system. Then the NF technology found its use in the efficient removal of dyes from textile wastewater, pharmaceutical separation or enrichment, heavy metal rejection, and treatment of organic solvent systems. To meet the requirements of different feed sources, membranes with diverse pore sizes, chemical structures and charge properties have been developed for a series of specific applications.

4.2 Membrane materials

Traditional NF membranes are made from cellulose acetate and thin film interfacial polymerization. In the last decade, Singapore researchers investigated the feasibility of using polybenzimidazole (PBI) and polyamide-imide (PAI) in the NF separation as well as supporting materials for the fabrication of NF membranes. Figure 7 compares the chemical structures of the four materials. Both PBI and PAI are rich

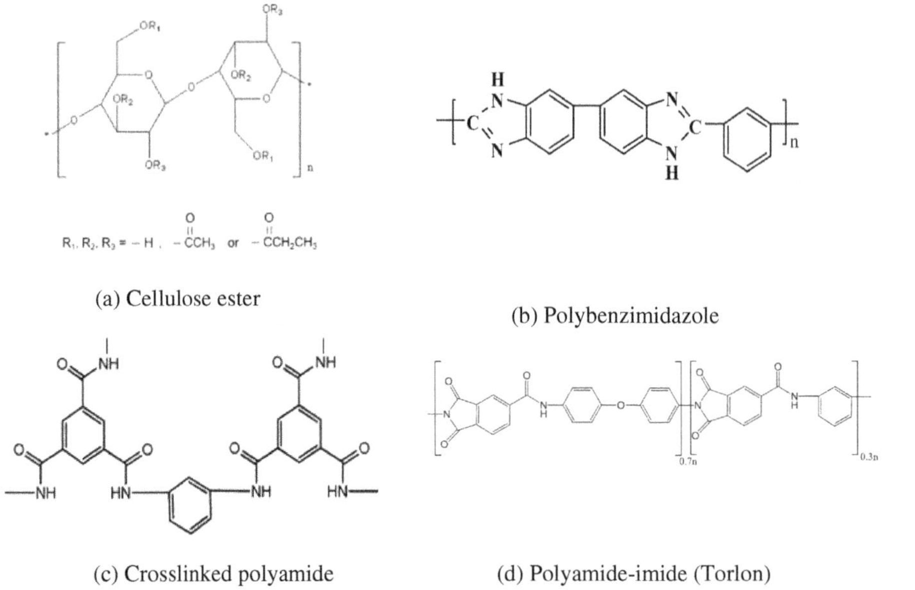

Fig. 7. The chemical structures of (a) cellulose ester, (b) polybenzimidazole, (c) crosslinked polyamide and (d) polyamide-imide (Torlon) as the base materials for selective layers of the reverse osmosis, nanofiltration or forward osmosis membranes.

in functional groups that can participate in subsequent crosslinking reactions. A variety of crosslinkers have been chosen to manipulate the pore structure and surface charge properties. Their shape, length and chemistry greatly influence the selective layer and the final separation performance. Figure 8 gives an example of the crosslinked structure of PAI by the hyperbranched polyethyleneimine (PEI) (Sun et al., 2011). The introduction of PEI not only reduces membrane pore size, but also enrich the surface with abundant free amine groups so that the resultant NF membrane exhibits positive charges in a wide range of pH.

Efforts have also been made to produce the ready-to-use membranes for NF rejections in one step. Ong and Chung (2014b) invented the incompatibility-induced phase separation (I²PS) technique to fabricate a dual layer polyacrylonitrile (PAN) — PAI (P84) hollow fiber. The rejection layer was formed at the P84 side near the interface of the two layers. Without any tedious post-modifications, the as-spun fiber shows a pure water permeability of 3.2 ± 0.5 L m^{-2} h^{-1} bar^{-1}, an MWCO of 470 Da, and a high rejection to SO_4^{-2} ions.

The thin NF selective layer can be also formed by other interfacial polymerization and UV crosslinking approaches. Hyperbranched PEI and isophthaloyl chloride were interfacially crosslinked to create a positively charged, relatively loose barrier layer (Sun et al., 2012). Dendrimers were attached to the surface of the interfacially formed polyamide layer to increase the surface charge (Zhu et al., 2015). In addition,

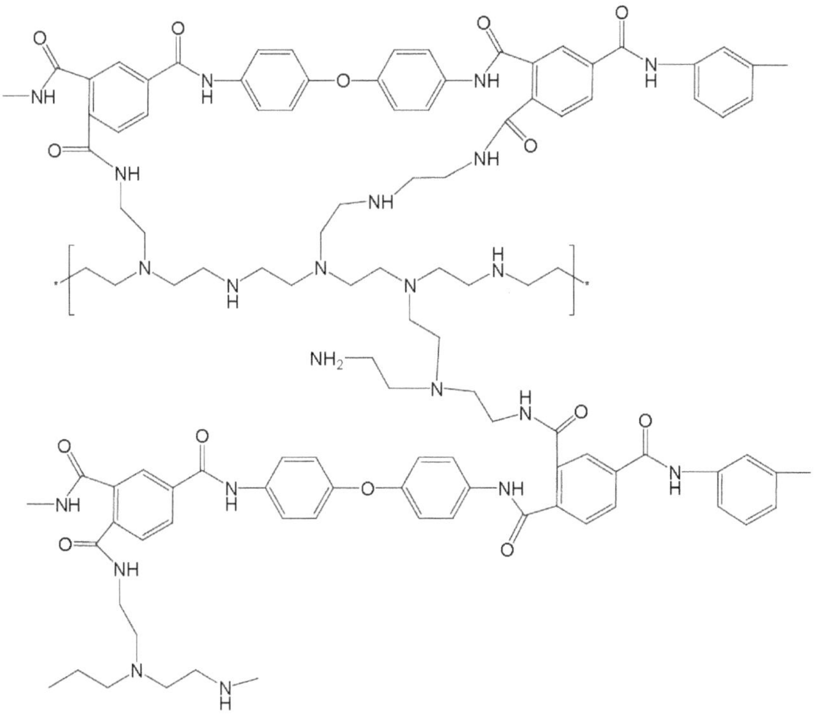

Fig. 8. An example of the crosslinked structure of polyamide-imide by hyperbranched polyethyleneimine (Sun et al. 2011).

positively charged monomers were coated onto an sPPSU support to form the rejection layer by UV (Zhong et al., 2012b). Meanwhile, sulfonated copolymers and the natural polymeric chitosan were employed to establish the NF selective layers (Thong et al., 2014, Zhang et al., 2014).

Local research interests have been focused on the design of NF membranes for heavy metal removal, recycling textile wastewater and organic solvents. The following section will introduce more details.

4.3 Nanofiltration membranes for textile wastewater treatment

The textile industry is one of the major industries in the world. It is also one of the most polluting industries. Depending on the process and textile materials, about 2–180 litres of wastewater are generated for each kilogram of textile products (International Finance Corporation, 2007). The wastewater is generally high in chemical oxygen demand (COD) and biochemical oxygen demand (BOD). Commercial NF membranes have been explored to treat the textile wastewater. However, the commercially available membranes are pre-dominating negatively charged TFC membranes with the flat-sheet configuration. They are not suitable to treat positively charged dyes because severe fouling is encountered when the textile wastewater consists of positively charged dyes.

To effectively remove positively charged dyes, NUS and BASF jointly fabricated positively charged NF membranes by UV grafting two positively charged hydrophilic monomers onto an sPPSU UF support (Zhong et al., 2012b). The sPPSU was chosen as the support because it would generate free radicals upon exposure to UV light (Yamagishi et al., 1995), which triggers the polymerization of pre-coated vinyl monomers on the surface. After the UV grafting, the resultant membranes have an overall positive charge under the testing conditions and are able to achieve a high rejection (>99%) of Safranin O (a positive charged dye with a molecular weight of 350.84Da) with a high pure water permeability (PWP) of 9–14 L m^{-2} h^{-1} bar^{-1}.

Alternatively, a double repulsive NF membrane was fabricated by the interfacial polymerization over the polyamide-imide support, in a work by NUS (Sun et al., 2011). Due to the unique combination of a positively charged rejecting layer and a negatively charged support, the membrane showed high rejections to both positively and negatively charged dyes. Ong et al. (2014a) demonstrated a pilot study of the hyperbranched PEI crosslinked polyamide-imide hollow fibers for the treatment of actual textile wastewater. A photo of the pilot system is provided in Fig. 9. The membranes exhibited satisfactory rejections (>90%) against various

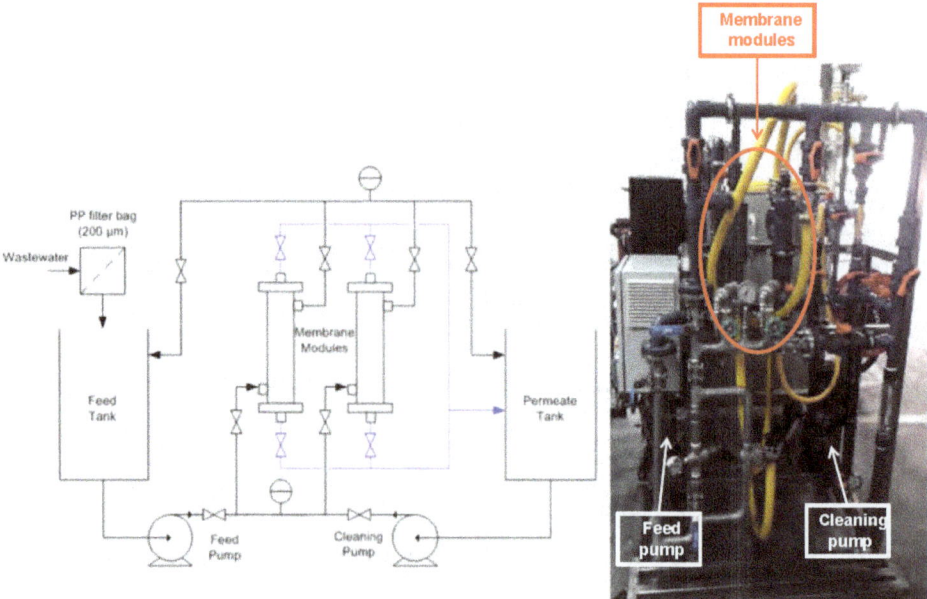

Fig. 9. Schematic of the pilot nanofiltration system for dye removal. Reprinted with permission from Ong et al., 2014a.

dyes with a water permeability of 6 L m^{-2} h^{-1} bar^{-1}. Moreover, on-site pilot studies over a 45-day period proved the robustness of the membranes in terms of both water flux and rejection.

4.4 Nanofiltration membranes for heavy metal removal

Heavy metal pollution has become a great concern in many countries including Singapore due to the indiscriminate discharge of untreated wastewater. Heavy metals are highly toxic, non-biodegradable and tend to be bio-accumulated in living organisms. NF provides a promising option. Due to the fact that most heavy metal ions are small and positively charged in aqueous solutions, membranes with small pore size and positive charges are preferred to achieve the simultaneous size and Donnan exclusions. Meanwhile, negatively charged membranes are also investigated.

Earlier NUS studies used PBI NF hollow fiber membranes to remove chromate (Wang et al., 2006). PBI possesses impressive chemical stability over a broad range of pH and thermal stability. Moreover, it is self-charged in aqueous solutions. Recently, a PEI modified polyamide-imide hollow fiber was developed by NUS to remove lead ions with a 91.05% rejection of Pb^{2+} ions (Gao et al., 2014). The performance can be further enhanced by reducing the extent of cross-linking and coating a second layer of negatively charged chelating polymers on the membrane surface in the way similar to the layer-by layer approach (Gao et al., 2015). The resultant membranes have a higher permeability with outstanding rejection levels (>98%) to numerous types of heavy metal ions such as Pb(NO$_3$)$_2$, CuSO$_4$, NiCl$_2$, CdCl$_2$, ZnCl$_2$, Na$_2$Cr$_2$O$_7$ and Na$_2$HAsO$_4$.

In contrast, a layer of slightly negative charged Nexar™ copolymer (Geise et al., 2010), which consists of pentablock copolymer, was deposited on top of the PEI modified polyamide-imide flat sheet substrate by Thong et al. (2014). The resultant membranes had rejection rates of greater than 98% for heavy metal cations such as Pb^{2+}, Cd^{2+}, Zn^{2+} and Ni^{2+}. Slightly lower rejections of 99.86% and 92.32% were reported for heavy metal anions HAsO$_4^{2-}$ and HCrO$_4^{-}$, respectively.

So far the highest rejections and water permeability were achieved by the dendrimer-modified polyamide thin film composite hollow fiber membranes (Zhu et al., 2015). In addition, a thin film interfacial crosslinking approach was adopted to improve the rejection of the chitosan layer towards heavy metal ions (Zhang et al., 2014).

4.5 Organic solvent nanofiltration (OSN) membranes

Although NF has been widely applied in the filtration of aqueous mixtures, its application towards organic solvent systems is still limited, mainly due to the lack of chemical resistance. To resist the swelling or even dissolution by the organic solvents, membranes with high degrees of crosslinking are usually required in the organic solvent nanofiltration process.

To this aim, NUS scientists prepared OSN membranes by chemically cross-linking PBI with either glutaraldehyde or diexpoxyoctane (Xing et al., 2014). The glutaraldehyde cross-linking method produced PBI membranes with high ethanol and ethyl acetate fluxes but with weak resistance in aggressive solvents such as dimethyl sulfoxide (DMSO). In contrast, PBI membranes cross-linked by DEO exhibited lower ethanol and ethyl acetate fluxes but had impressive stability in aggressive solvents such as DMSO, N-methyl-2-pyrrolidinone (NMP) and N,N-dimethylacetamide (DMAc). Overall, the cross-linked PBI membranes exhibited similar rejections to different solutes but the rejections differs slightly from solvent to solvent due to different solute–solvent–membrane interactions.

5. Membrane distillation

5.1 Mechanism

Membrane distillation (MD) is an emerging technology that combines membrane contactor and evaporation processes. As illustrated in Fig. 10, the hydrophobic nature of the membrane pores resists the liquid penetration. Meanwhile, volatile components in the hot feed stream vaporize and are transported across the membrane due to vapor pressure gradients. The vapors are then collected by different methods.

MD offers a number of advantages compared to other membrane processes: (1) operation permissible at atmosphere pressure and relatively low temperatures (30–90°C); (2) theoretical ability to achieve 100% rejection of non-volatile ionic solutes and macromolecules; and (3) less sensitive to the solute concentration in the feed, which makes it especially attractive for the desalination of high salinity water

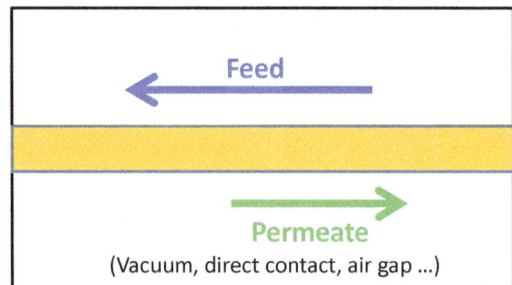

Fig. 10. An illustration of the membrane distillation process.

(Lawson and Lloyd, 1997; Curcio and Drioli, 2005). If low-cost solar and geothermal energy, or industrial waste heat are available, it may be able to compete with RO for desalination in terms of cost and energy efficiency (Burgoyne and Vahdati, 2000). In addition, MD is also proposed for other applications such as wastewater treatment and oil water separation.

In Singapore, a MEMSTILL® MD pilot plant was tested for desalination in 2010 by Keppel Integrated Engineering Ltd in the Senoko Refuse Incineration Plant. (PUB, 2010). Keppel also collaborated with MEMSTILL on MD for seawater desalination in its platforms. (Keppel, 2011)

However, full commercialization of the MD technology worldwide is constrained by two factors: (1) the lack of high-performance commercialized MD membranes and (2) the availability of heat sources. Design of suitable membranes for MD has been the major research focus in the past years. The desired membranes should possess high flux, low thermal conductivity, excellent anti-wetting properties and resistance towards high temperature, fouling, and scaling. As thermal and mass transport co-exist in the process, membranes with high porosity and low thermal conductivity are needed to reduce mass diffusion resistance and heat loss. Wetting refers to the entrance of liquids into the membrane pores. It happens during continuous MD operations and deteriorates the rejection to solutes. Better wetting resistance requires higher hydrophobicity and uniformed distributed small pores (Lawson and Lloyd, 1997; Edwie et al., 2012).

An introduction to the progress in MD membranes will be given in the following, starting from the membrane materials first and then the three generations of membranes that have been developed in Singapore.

5.2 Membrane materials

Hydrophobicity is an essential requirement for MD membranes in most applications. Materials with low surface energy are usually chosen for membrane fabrication and modification. Typical commercially available polymeric materials for MD are listed in Table 2 (Wang and Chung, 2015). So far, the most popular polymers are polytetrafluorethylene (PTFE), polyvinylidene fluoride (PVDF) and polypropylene (PP).

PTFE and PP are highly crystalline polymers with low surface energy. PTFE membranes are most frequently used in the pilot MD systems because of their excellent stability in diverse operating conditions, good wetting resistance and satisfactory water flux. Due to their low solubility in many organic solvents, it is difficult to fabricate membranes based on PTFE or PP by the common nonsolvent-induced-phase-inversion (NIPS) process. Instead, their membranes are normally prepared by sintering, melt-extrusion and melt-extrusion-stretching methods.

Table 2. Characteristic Properties of Commercial Polymer Materials Commonly used for MD Membranes. Reprinted with Permission from (Wang and Chung, 2015).

Polymer materials	Chemical structure	Surface energy ($\times 10^{-3}$ N m^{-1})	Thermal conductivity (W m^{-1} K^{-1})	Thermal stability	Chemical stability	Fabrication methods
PTFE		9–20	0.25	Good	Good	Sintering Melt-extrusion
PP		30	0.17	Moderate	Good	Melt-extrusion TIPS
PE		28–33	0.40	Poor	Good	Melt-extrusion TIPS
PVDF		30.3	0.19	Moderate	Good	NIPS TIPS Electro-spinning
PVDF-HFP		—	—	Good	Good	NIPS Electro-spinning
Hyflon®		—	0.2	Good	Good	NIPS

PVDF is a semi-crystalline polymer with a relatively low melting temperature of 170°C and can be easily dissolved in common solvents. Therefore, PVDF membranes can be prepared by either NIPS, temperature-induced-phase-inversion (TIPS) or electro-spinning processes.

Apart from polymers, metal, glass, carbon nanotubes (CNT), and other inorganic based materials have also been evaluated for MD. In addition, additives and modifications to the polymers have been conducted to alter the membrane's properties. Small molecules, inorganic salts, particles or macromolecules were added to manipulate the pore size and hydrophobicity; plasma was also applied to induce chemical modifications. However, for the applications of MD in oil-like liquid separation, hydrophilic materials such as polyacrylonitrile (PAN) were preferred instead of hydrophobic PVDF (Qu et al., 2008). Membranes made from PVDF were found to suffer from serious wetting by the oil in the separation of petroleum ether from the solanesol extraction solution.

5.3 Membrane development

Chemical and structural properties are critical for MD membranes to achieve high water flux and good stability. Besides materials, the macro-geometry and micro-structure of the membranes play important roles in membrane performance. Membranes with large pores in the range of 0.1 –0.3 µm and a high porosity are preferred to facilitate vapor transport and minimize heat loss, on the basis of minimal wetting at the pores. At least three generations of MD membranes have been developed by Singapore scientists.

The first generation MD hollow fiber membranes developed by Singapore were single-layer membranes. To increase the membranes' porosity, ethylene glycol (EG) was added into the spinning dopes (Wang et al., 2008). With its low molecular weight and good miscibility with water, EG can be readily leached out from the nascent membrane after spinning. The as-spun fiber demonstrated a flux enhancement of about 4.5-fold. Next, a co-extrusion method was developed to introduce high surface porosity in the support (Bonyadi and Chung, 2009). In this approach, the polymer solution and solvent are co-extruded from the middle and outer channels of a triple orifice spinneret, respectively, before entering the coagulant bath. A macrovoid-free, sponge-like and open-cell structure had resulted from the delayed demixng induced by the outer-channel solvent.

Mixed matrix membranes consisting of micro- and/or nano-particles in the polymer phase provide a promising approach to overcome deficiencies in physicochemical properties of membrane materials (Chung et al., 2007). By incorporating different particles, one can alter the membrane morphology and chemistry.

Incorporating super-hydrophobic PTFE particles into the PVDF matrix was found to not only significantly enhance the outer surface hydrophobicity (Teoh and Chung, 2009), but also effectively suppress the macrovoid formation. Hydrophobic clay particles (Cloisite 20A) were also employed to form composite PVDF-Clay hollow fibers with enhanced mechanical strength and insulation properties (Wang et al., 2009a). In another work (Edwie et al., 2012), fluorinated silica particles were blended with PVDF solutions to form highly hydrophobic membranes with a water contact angle of 147.5°.

In the second generation, the state-of-the-art dual-layer spinning technique was applied to independently control the rejecting and support layer properties. Two types of fibers have been produced — namely, hydrophilic/hydrophobic dual-layer hollow fibers and dual-morphology membranes. The hydrophilic-hydrophobic dual layer concept was first proposed in the early 90's (Khayet and Matsuura, 2003). The thin hydrophobic rejecting layer prevents the wetting by water and the hydrophilic support layer reduces the vapor transport resistance. In the recent NUS work (Bonyadi et al., 2008), hydrophobic clay particles were added in the outer PVDF layer to achieve a water contact angle of 137°, and hydrophilic clay particles were blended with polyacrylonitrile (PAN) and PVDF to form the inner layer. PVDF was used in both layers to improve molecular inter-diffusion so that delamination between the two layers would not occur.

To reduce the mass transfer resistance in the support layer, a dual-layer hollow fiber membrane consisting of a sponge-like outer layer and a fully finger-like macrovoid inner layer was invented by NUS (Wang et al., 2011). The outer layer functions as the rejecting layer to resist wetting, while the finger-like macrovoids could minimize the tortuosity and facilitate vapor transfer. A superior water flux of 98.6 L m^{-2} h^{-1} was achieved for seawater desalination at the feed inlet temperature of 80°C. The energy efficiency reached as high as 94%.

In order to enhance the fiber's mechanical strength in both axial and radial directions, the third generation membranes — namely, multi-bore hollow fibers were designed and developed (Teoh et al., 2011; Wang and Chung, 2014a). The geometry of the fibers are similar to the ultrafiltration membranes as shown in Fig. 5. In addition to high permeation flux, superior stability and robustness in the long term operations were demonstrated.

The three generations of MD membranes were all based on the hollow fiber technology. Other efforts have also been made to enhance membranes' performances. Electro-spun nanofibers have been employed to fabricate PVDF nanofiber membranes at NTU for MD (Liao et al., 2013a). Later, surface modification by dopamine activation, silver nanoparticle deposition and hydrophobic treatment was applied to obtain superhydrophobic PVDF membranes (Liao et al., 2013b).

6. Forward Osmosis

6.1 The osmosis process

Osmosis is a physical phenomenon that has been extensively explored since the 19th century. When two solutions of different concentrations are separated by a semipermeable membrane, i.e., permeable to the solvent but impermeable to the solute, water flows naturally from the high chemical potential side to the lower side until an equilibrium is reached. The increased volume of water in the low chemical potential side builds up a hydrodynamic pressure difference, which is called the osmotic pressure difference $\Delta \pi$. By utilizing a high concentration solution as the draw solution and a low concentration as the feed, effective separation of water from the feed can be achieved across the membrane. This process is termed as forward osmosis (FO).

The driving force in FO is the osmotic pressure difference. This is different from the conventional RO process, where a hydraulic pressure higher than the osmotic pressure of the feed is applied ($\Delta p > \Delta \pi$), and water is pressed out of the feed and collected as the permeate. When a moderate hydraulic pressure is applied upon the feed but is still lower than the osmotic pressure ($\Delta p < \Delta \pi$), it is called the pressure retard osmosis (PRO) process. The permeated water is pressurized and can be used to generate power. A comparison of the three processes is illustrated in Fig. 11 (Alsvik and Hägg, 2013).

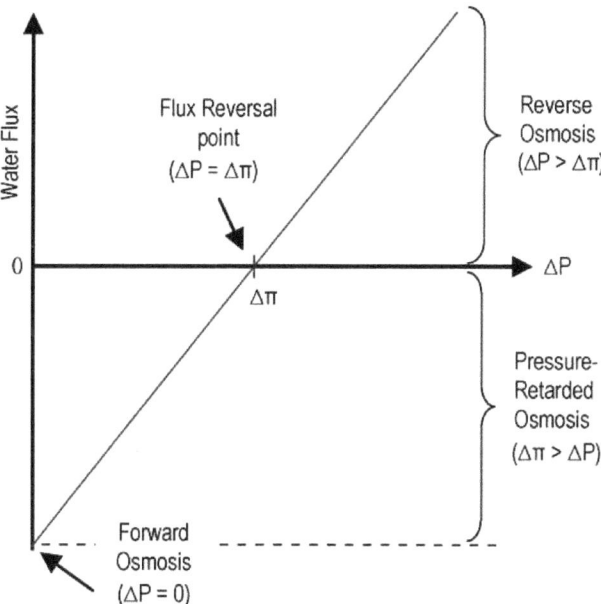

Fig. 11. Direction and magnitude of water flux as a function of applied pressure in FO, PRO, and RO. Reprinted with permission from Elsevier (Lee et al. 1981).

Compared with traditional pressure-driven processes, FO features a few advantages: (1) direct applications in the dilution of concentrated solutions such as juices and sucrose; (2) mild operation conditions for pressure-sensitive solutes; (3) possible combination with MD and others for draw solution recovery and water production; and (4) high rejections towards a wide range of molecules. It can have wide applications in various fields, e.g. desalination, agricultural irrigation, protein concentration, oil water separation, wastewater treatment and many others.

6.2 Required FO membrane characteristics

To maximize the water flux and minimize the salt reverse flux in the FO applications, membranes must be designed with the following characteristics:

(1) A thin and dense selective layer to reject the solutes in both the draw and feed solutions. Depending on solution compositions and applications, one may require different membrane selectivity. If sodium chloride is involved, the dense layer should resemble RO membranes with a very low pore size of around 0.3 Å. So far limited materials have been used in this category, mainly including cellulose acetate (CA)/cellulose triacetate (CTA) and crosslinked polyamide formed by interfacial polymerization. On the other hand, if the solutions consist of larger molecules, a looser structure in the selective layer is acceptable as long as the solutes can be effectively rejected. Some materials have been utilized for such purposes, e.g., PBI, crosslinked polyamide-imide (PAI) and layer-by-layer (LbL) polyelectrolytes.

(2) A thin and highly porous support layer to minimize the ICP effects. High hydrophilicity is also needed to improve the wetting properties in the support (Widjojo et al., 2013). In general, NUS scientists have found that relatively hydrophilic materials are preferred to construct the support, such as polyethersulfone (PES), sulfonated polymers, cellulose derivatives, PAN, etc. Chemical modification with polydopamine (PDA) is another approach to make the support more hydrophilic. The structure of the most frequently used materials for the selective layer can be found in Fig. 7.

Singapore is among the pioneers to work on FO. Research work has been very active to fabricate proper membranes for FO and to investigate its potential applications in different areas. Some of them are introduced in the next sections.

6.3 Cellulose ester and thin film composite (TFC) polyamide FO membranes

CTA and TFC polyamide membranes have been widely applied in the traditional RO desalination processes. Nonetheless, these RO membranes perform badly in the FO process due to their thick support layer. Innovations in FO membrane development mainly start from the design of a proper structure to reduce mass transfer resistance. In addition, new cellulose derivative polymers have been synthesized and made into FO membranes. In the case of TFC polyamide membranes, the support surface properties affect the formation of the thin film layer. Such interplay has also been studied and utilized to improve the overall membrane performance.

Cellulose ester polymers can be easily synthesized from the chemical substitution of cellulose. They offer advantages which include vast availability, low costs, hydrophilicity, tunable chemical structure, ease of fabrication and chlorine resistance. HTI was the first to develop and manufacture CTA-based FO membranes with woven supports in the market. The woven support is only 50 μm thick and relatively porous. The membranes achieve superior FO performance than RO membranes.

Hollow fiber FO membranes had been first developed by NUS using CA (Su et al., 2012) and PBI (Wang et al., 2007). Annealing was found necessary to increase the rejection to salts. Double-skin flat sheet membranes were also firstly produced by NUS (Wang et al., 2010; Zhang et al., 2010). This concept will be touched on in Section 2.6. By tuning the content of various functional groups in the cellulose ester, the FO performance of the resultant membranes can change significantly (Zhang et al., 2011; Ong and Chung, 2012a; Ong et al., 2012b). FO membranes made from cellulose esters with a very high hydroxyl content tend to have high water and solute permeability. A moderate content of hydroxyl and propionyl or butyryl groups is preferred to achieve a reasonable FO water flux and solute rejection. Considering the good permeation properties and relative hydrophobicity of cellulose acetate propionyl (CAP) polymers, dual layer hollow fibers made of a thin CAP selective skin layer on a porous CA layer were produced and it demonstrated high performance in the FO process (Su et al., 2013).

The TFC membrane for FO was first developed by NTU and Yale scientists following RO's interfacial polymerization on top of a porous support (Yip et al., 2010; Wang R et al., 2010). Typically, m-phenylenediamine (MPD) in water and trimesoyl chloride (TMC) in hexane were used. The interfacial polymerization takes place extremely fast (even less than one minute) at the organic phase near the interface, leading to the formation of a dense polyamide thin film layer on top of the support. Comparing to cellulose ester materials, the crosslinked polyamide produces a higher intrinsic water permeability and a comparable salt rejection, but its chlorine and fouling resistance are lower.

It was found by NUS scientists that a fully sponge-like macrovoid-free and hydrophilic support may provide TFC FO membranes with better balance between FO performance and mechanical properties (Sukitpaneenit and Chung, 2012; Widjojo

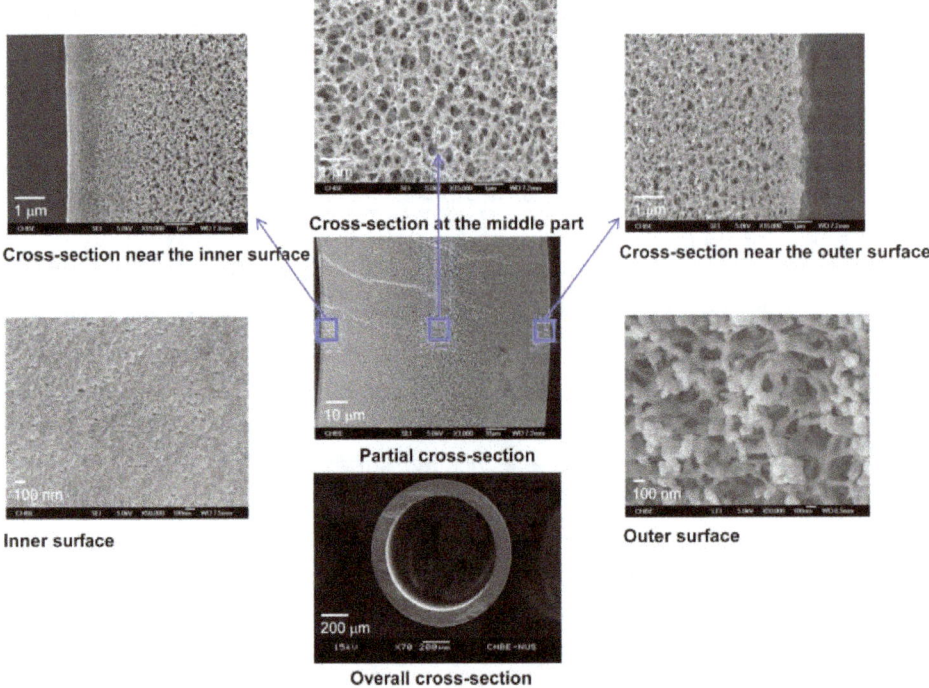

Fig. 12. SEM images of the hollow fiber support for thin film composite membranes in forward osmosis. Reprinted with permission from (Sukitpaneenit and Chung, 2012).

et al., 2013; Zhong *et al.*, 2013). A new approach using a dual-layer co-extrusion technology was proposed to fabricate macrovoid-free and highly porous sponge-like PES hollow fiber substrate as illustrated in Fig. 12. Several methods were combined to produce the desirable structure, including the addition of a pore-forming agent, such as polyethylene glycol 400, and a non-solvent, such as water, and the flowing of a solvent at the outer channel of the spinneret. The resultant fibers showed a superior water flux among this type of membranes while maintaining a reasonable salt rejection.

To enhance the wetting properties in the support and ultimately improve the membrane performance, hydrophilic polymers, such as cellulose esters and sulfonated polymers were studied by NUS (Li X *et al.*, 2012a; Widjojo *et al.*, 2013; Zhong *et al.*, 2013). The research was conducted in close cooperation with local and international companies, e.g. Eastman Chemical Company, BASF, Mitsui Chemicals, who provided the newly synthesized polymers. Additionally, hydrophilic modifications by polydopamine were done to the membrane support (Han *et al.*, 2012), which not only improved the formation of the thin film layer, but also reduced the ICP effects.

An alternative concept of nanofiber support with high porosity and low tortuosity characteristics by means of an electrospinning technique has also been applied.

Song et al. (2011) at NTU prepared the polymeric nanofiber supports and found a very small gap between PRO and FO performance, indicating a remarkably reduced ICP. It is mainly attributed to the unique scaffold-like porous structure and highly interconnected pores of the nanofiber supports.

6.4 Biomimetic membranes

Nature provides humans with inspiration and answers to problems throughout our existence. Biomimetic membranes exhibit outstanding desalination performance for practical use by mimicking the natural biological membranes, which exhibit high water permeability and excellent selectivity towards salts due to the presence of the transmembrane proteins, known as aquaporins (AQPs). Early efforts involved the fabrication of black lipid membranes, vesicles and solid-supported lipid/block copolymer membranes. Kaufman et al. (2010) designed a solid-supported biomimetic membrane, where a biomimetic lipid layer was coated on the surface of a commercial nanofiltration membrane.

The major challenge for the development of biomimetic membranes is the incorporation of Aquaporins without damaging their functionality. Typically a two-step procedure is adopted. Preparation of aquaporin-incorporated vesicles is the first essential step and is then followed by the deposition of vesicles on a solid support. Lipids or block copolymers are used to form the vesicles by four different techniques, including mechanical ways, organic solvent-mediated incorporation, direct incorporation into preformed vesicles and detergent-mediated reconstitution (Rigaud et al., 1995). The detergent-mediated method has proven to be the most successful means and is the method most frequently used.

In the second step, the supported lipid/block copolymer layer is formed by the spontaneous deposition of vesicles on a solid support. A general principle is that vesicle ruptures will induce a sufficiently large deformation which will then transform onto a flat bilayer. These rupture events are governed by the interaction between the membrane and the support as well as intra-membrane interactions, such as electrostatic interactions and covalent bonding. The major difficulties in the fabrication of biomimetic membranes are to ensure sufficient mechanical stability and eliminate defects between the selective layer and the solid support, while maintaining the functionality of aquaporins.

Research on biomimetic membranes became very active in Singapore since 2008. Chung's group in NUS started to investigate the fabrication and application of biomimetic membranes for FO (Wang H et al., 2012; Wang H et al., 2013). Joint research among NTU, Danish company Aquaporin A/S and DHI Water & Environment had aimed to develop highly permeable membranes for pressure-driven desalination processes (PUB, 2013; Li X et al., 2012b; Zhao et al., 2012).

Innovative membranes have been developed. Nanoporous membranes were used to provide mechanical support and anchoring points. Aquaporin-containing polymersomes were ruptured and crosslinked on methacrylate-functionalized cellulose acetate membranes (Zhong et al., 2012a). Charged lipid vesicles were deposited and stabilized on negatively charged nanofiltration membranes at pH 7 (Li X et al., 2012b). Polymersomes were also deposited on the acrylate groups functionalized polycarbonate track-etched substrates (Wang H et al., 2012). However, despite of high permeability, the long term stability of these structures is a concern for industrial applications.

By coating or embedding aquaporin-containing proteoliposomes or proteo-polymersome in a crosslinked polyamide matrix (Wang H et al., 2013) a robust membrane structure has been achieved. An illustration is shown in Fig. 13. The membranes showed great potential for FO desalination by demonstrating water fluxes of up to 142 L m^{-1} h^{-1} when 2.0 M NaCl was used as the draw solution (Wang H et al., 2013). A comparison of the representative membrane performances over the past decade is provided in Table 3.

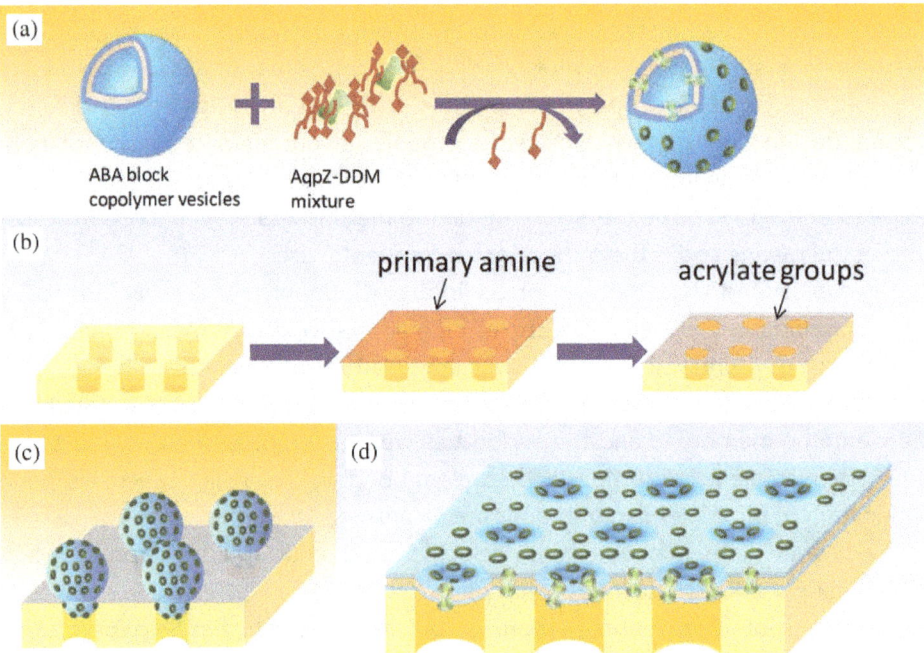

Fig. 13. Schematic diagram of pore-spanning biomimetic membrane synthesis procedures: (a) Incorporation of AqpZ in ABA block copolymer vesicles; (b) Surface modification of the PCTE membrane support. (c) Pressure-assisted vesicle adsorption on the support. (d) Covalent conjugation-driven vesicle rupture and pore-spanning membrane formation. Reprinted with permission from (Wang H et al., 2013).

Table 3. FO Performances of the Representative Membranes. DI Water was Used as the Feed and Experiments were Conducted at 20–25°C.

Configuration	Material (selective layer)	Water Flux, L m^{-2} h^{-1} (FO/PRO)	Reverse Salt flux, g m^{-2} h^{-1} (FO/PRO)	Draw Solution	Reference
Double-skin flat sheet	Cellulose acetate	22/32	16/25	2.0 M NaCl	Wang KY et al., 2010
Dual layer hollow fiber	Cellulose acetate propionate	8.5/17.5	1/2	2.0 M NaCl	Su et al., 2013
Dual layer hollow fiber	Polybenzimidazole	12/32	3/6	3.0 M MgCl$_2$	Fu et al., 2013
Dual layer hollow fiber	Polyamide-imide	15.5/27.5	5.5/83.7	0.5 M MgCl$_2$	Setiawan et al., 2012
Hollow fiber	Crosslinked polyamide	18.0/37.7	2.6/7.0	0.5 M NaCl	Zhong et al., 2013
Flat sheet	Crosslinked polyamide	62.8/78.8	14.9/13.5	2.0 M NaCl	Natalia et al., 2013
Flat sheet	Polyelectrolytes (LbL)	11.6/11.8	8.1/13.0	0.5 M NaCl	Duong et al., 2013
Flat sheet	Biomimetic	142 (PRO)	6.2	2.0 M NaCl	Wang H et al., 2012

6.5 Double-skin FO membranes

Due to the existence of a porous support in the membranes, foulants from the feed solution may easily block the pores and reduce the water flux in the PRO mode where the draw solution faces the dense selective layer; similarly in the FO mode where the feed solution faces the dense selective layer, viscous draw solutes may also experience hindered diffusion and pore blockage, thus decreasing the membrane performance as a combined result of ICP and fouling issues.

NUS was the first to develop double-skin membranes for FO (Wang et al., 2010). Zhang et al. found that by controlling the polymer and substrate affinity during the casting of flat sheet membranes, a dense selective layer could be formed at the polymer-substrate interface, and a second looser skin layer at the polymer-air interface demonstrated low and reversible fouling towards modeled nanoparticle foulants (Zhang et al., 2010; Zhang et al., 2011). Fig. 14 illustrates a comparison of the anti-fouling behaviors of the double- and single-skin membranes. Fang et al. (2012) at NTU also noticed the improved anti-fouling behavior by the dual selective hollow fiber membranes consisting of dense RO-like and slightly looser NF-like selective layers. Besides, dual selective flat sheet membranes based on LbL techniques were produced by Qi et al. (2012). Good anti-fouling performance was observed by using Dextran (Mw 200–300 kDa) and alginate acid sodium salt

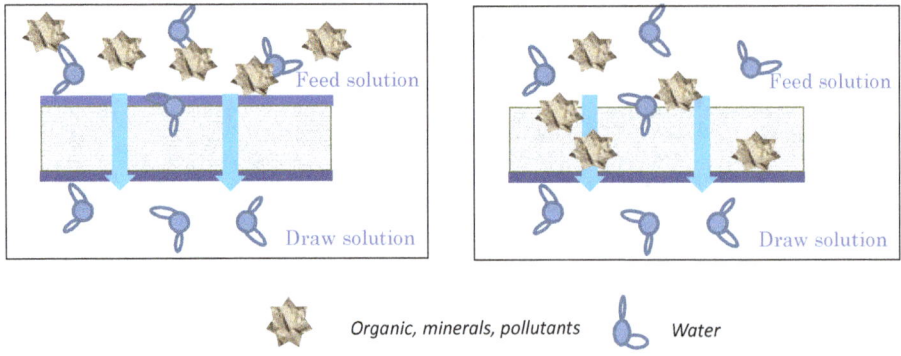

Fig. 14. Illustrative comparisons of the anti-fouling behaviors of (left) double- and (right) single-skin membranes in the PRO mode.

(Mw~12–80 kDa) as the model foulants. In a more recent NUS work, sulfonated materials were coated on the other surface of a TFC membrane to resist the severe fouling by emulsified oils (Duong et al., 2014).

The Singapore work showed that double-skin membranes can be designed with different selectivity on each side by altering the preparation methods. This might provide a good opportunity to tune the membrane structure for specific applications. Depending on the feed sources, foulants may have different sizes and electrochemical characteristics. By carefully designing double-skin membranes with different surfaces on each side, one can maximize the water flux and minimize the fouling effects in specific FO applications.

6.6 Pressure retarded osmosis membranes

A special application of FO is for osmotic power generation via pressure retarded osmosis (PRO). By applying a hydraulic pressure ΔP ($\Delta P < \Delta \pi$) to the draw solution, water that permeates across the membrane from the feed is then pressurized and can be used to generate power. PRO was conceptualized for harvesting salinity gradient energy by Prattle (Prattle, 1954) in the 1950s and was first run by Loeb et al. in the 1970s (Loeb et al., 1976). In 2009, the first PRO prototype plant was built in Norway which demonstrated the generation of electricity by the PRO process (Skilhagen, 2010). However, a poor power density of lower than 3 W m^{-2} was observed due to the lack of effective membranes.

Different from other FO membranes, the desirable PRO membranes should possess not only a high water permeability and a reasonably low salt permeability, but also good mechanical strength without sacrificing the structural parameter. Rapid progresses has been made in Singapore in recent years, mainly through (1) increasing the water permeability of the selective layer by the modified membrane formation process and novel post-treatment methods and (2) improving the structure of the support layer to ensure a low structural parameter and good mechanical strength. Currently, the best PRO hollow fiber has a maximum power density of 24–27 W m^{-2} at 20 bar utilizing 1 M NaCl and DI water as feeds. This PRO hollow fiber consists of an ultrathin TFC layer and a PES porous support (Zhang and Chung, 2013; Wan and Chung, 2015).

The first high performance integrally skinned PRO membrane was the PBI-PAN dual layer hollow fiber made from direct phase inversion at NUS (Fu et al., 2014). PBI formed a dense skin at the outer layer and the PAN was the support. PVP was added into the PAN polymer dope as an effective bridge to avoid the delamination between the two layers. Later, extra PVP molecules were removed by the post-treatment of ammonium persulfate, resulting in a higher water permeability and power density.

The TFC membranes have higher PRO performance than the integrally skinned ones. Both flat sheet and hollow fiber configurations have been developed in Singapore for the TFC PRO membranes. To increase water permeability, various additives such as bulky monomers (Li X and Chung, 2013) and surfactants (Cui et al., 2014) were added during the interfacial polymerization. Post-treatment by chlorine treatment (Han et al., 2013a) and alcohol immersion (Zhang et al., 2013) were also found effective. In addition, super porous electron-spun nanofibers were employed as the support to minimize the structural parameter (Song et al., 2013).

The mechanical stability of TFC membranes is maintained by selecting intrinsically strong materials and microstructural control for the support. Matrimid®, polyetherimide, polyethersulfone and P84 co-polyimide which have rigid benzene backbones were used (Chou et al., 2013; Han et al., 2013b; Li X and Chung, 2013; Zhang and Chung, 2013; Sun and Chung, 2013). By designing a highly asymmetric support structure, which is fully porous in the bulk but dense near the selective layer, the resultant TFC fibers can withstand high hydraulic pressures.

The PRO fibers were tested at NUS using RO retentate as the draw solution and NEWater retentate as the feed (Wan and Chung, 2015), both of which came from local water treatment plants. However, serious fouling was encountered in the support layer. Anti-fouling membranes were hence developed by covalently binding superhydrophilic hyperbranched glycol (HPG) into the support (Li X et al., 2014; Cai et al., 2016). Fig. 15 illustrates the procedures. Reduced fouling and high flux recovery after cleaning were achieved.

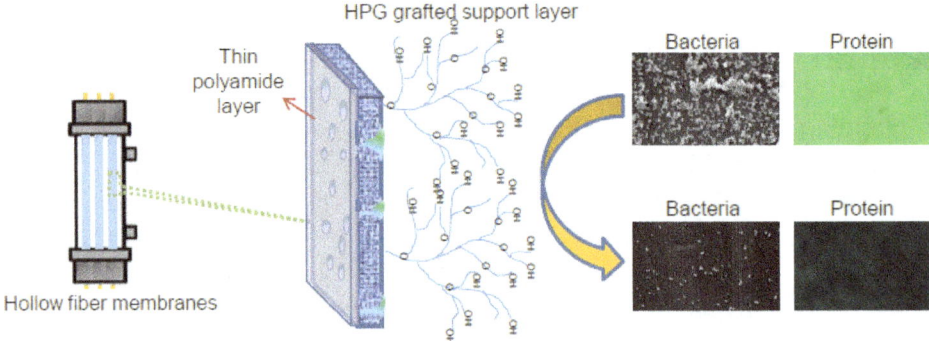

Fig. 15. Illustration of the grafting of hyper-branched polymer on the PRO membrane. Reprinted with permission from (Li et al., 2014). Copyright (2014) American Chemical Society.

6.7 Other FO and PRO membranes

As the salt rejection of the as-spun PBI layer was unsatisfactory, crosslinking by p-xylene dichloride on PBI fibers was carried out at NUS to tailor the pore size and enhance the selectivity for FO membranes (Wang et al., 2009b). Dual-layer FO hollow fibers were also produced by employing a supporting substrate, like polyethersulfone or polyacrylonitrile, and a thin PBI layer (Yang et al., 2009; Fu et al., 2013). In addition, polyvinylpyrrolidone (PVP) was added in polymer dopes to bridge the two layers to avoid delamination (Fu et al., 2013).

Polyamide-imide (PAI) was also employed by Setiawan et al. (2011) at NTU as the base material for FO membranes. The nascent PAI fibers were crosslinked by PEI molecules to reduce the pore size of the selective layer. The resultant membrane showed a poor rejection to NaCl, but reasonably good rejections to divalent ions. Since the PEI-modified PAI surface is positively charged, a negatively charged polystyrene sulfonate sodium salt (PSS) was deposited on top of the PEI layer to alter the surface electrochemical properties (Setiawan et al., 2012). Additionally, similar to the aforementioned PBI materials, dual-layer hollow fibers with polyethersulfone as the supporting layer have also been developed (Setiawan et al., 2012).

In addition, the LbL technique was utilized to produce charged polyelectrolyte surfaces for FO applications. Typically, polymer multilayers are assembled onto a charged membrane by the sequential adsorption of polyanions and polycations via dip-coating, spin-coating, or spraying methods. The structure is stabilized by electrostatic interactions (Decher G., 1997). The first few generations of LbL FO membranes developed by NTU showed reasonably good rejections to $MgCl_2$ with high water fluxes up to 100 LMH using a 2 M $MgCl_2$ as the draw solution (Qiu et al., 2012). Recently, Duong et al. at NUS crosslinked the polyelectrolyte layers to produce a relatively good rejection to NaCl (Duong et al., 2013).

7. Conclusion and Perspectives

Significant progress on membrane research and technology in Singapore has been made in the last 50 years. 25–30 years ago, we imported all the membranes from overseas, while today Singapore's has perfected on various membrane technologies and exports MF and UF membranes globally. In addition, Singapore's research on membranes for clean water and clean energy has received worldwide recognition. The Lux report from Lux Research USA, the world leading advisory firm providing strategic advice on research intelligence and emerging technologies, ranked NUS and NTU as the World's best two on water research in 2013. The success is mainly due to the combined efforts, strong Governmental support, and the society's dedication to overcome the water scarcity issue. However, the current RO process is energy intensive. Future membrane research should focus on (1) membranes with low energy consumption for water reuse and seawater desalination and (2) membranes for organic solvent recovery. The former is mainly for Singapore because we are an island country without sufficient groundwater, while the latter is to recover valuable organic solvents from our pharmaceutical and petrochemical industrial waste to lower CO_2 emission.

References

Alsvik, I.L., and M.B. Hägg, "Pressure retarded osmosis and forward osmosis membranes: materials and methods." *Polymers*, 2013. 5, 303–327.

Baker, R.W., "Membrane technology and applications." Wiley, California.

Bonyadi, S., and T.S. Chung, "Highly porous and macrovoid-free PVDF hollow fiber membranes for membrane distillation by solvent-dope solution co-extrusion approach." *J. Membr. Sci.*, 2009. 331, 66.

Bonyadi, S., K.Y. Wang, and T.S. Chung, "Fabrication of high performance dual layer hydrophilic-hydrophobic hollow fibers for membrane distillation process." US Provisional Patent Application, 2008. No. 61/193,359.

Bu-Rashid, K.A., and W. Czolkoss, "Pilot tests of multibore UF membrane at Addur SWRO desalination plant, Bahrain." *Desalination*, 2007. 203(1): pp. 229–242.

Burgoyne, A., and M.M. Vahdati, "Review. Direct contact membrane distillation." *Sep. Sci. Technol.*, 2000. 35, 1257.

Cadotte, J. "Interfacially synthesized reverse osmosis membranes." US patent 4277344, 1981.

Cai, T., X. Li, C.F. Wan, and T.S. Chung, "Zwitterionic Polymers grafted poly(ether sulfone) hollow fiber membranes and their antifouling behaviors for osmotic power generation." *J. Membr. Sci.*, 2016. 497, 142–152.

Chou, S., R. Wang, and A.G. Fane, "Robust and high performance hollow fiber membranes for energy harvesting from salinity gradients by pressure retarded osmosis." *J. Membr. Sci.*, 2013. 448, 44–54.

Chung, T.S., L.Y. Jiang, Y. Li, and S. Kulprathipanj, "Mixed matrix membranes (MMMs) comprising organic polymers with dispersed inorganic fillers for gas separation." *Prog. Polym. Sci.*, 2007. 32, 483–507.

Curcio, E., and E. Drioli, "Membrane distillation and related operation — a review." *Sep. Purif. Rev.*, 2005. 34, 35–86.

Decher, G., "Fuzzy nanoassemblies: Toward layered polymeric multicomposites." *Science*, 1997. 277, 1232–1237.

Duong, P.H.H., J. Zuo, and T.S. Chung, "Highly crosslinked layer-by-layer polyelectrolyte FO membranes: Understanding effects of salt concentration and deposition time on FO performance." *J. Membr. Sci.*, 2013. 427, 411–421.

Duong, P.H.H., T.S. Chung, S. Wei, and L. Irish, "Highly permeable double-skinned forward osmosis membranes for anti-fouling in the emulsified oil-water separation process." *Environ. Sci. Technol.*, 2014. 48, 4537–4545.

Edwie, F., M.M. Teoh, and T.S. Chung, "Effects of additives on dual-layer hydrophobic–hydrophilic PVDF hollow fiber membranes for membrane distillation and continuous performance." *Chem. Eng. Sci.*, 2012. 68, 567–578.

Eriksson, P., "Nanofiltration extends the range of membrane filtration." *Environ. Prog.*, 1988. 7(1): pp. 58–62.

Fang, W., R. Wang, S. Chou, L. Setiawan, and A.G. Fane, "Composite forward osmosis hollow fiber membranes: Integration of RO- and NF-like selective layers to enhance membrane properties of anti-scaling and anti-internal concentration polarization." *J. Membr. Sci.*, 2012. 394–395, 140–150.

Fane, A. G., R. Wang, and M.X. Hu, "Synthetic Membranes for Water Purification: Status and Future." *Angew. Chem. Int. Ed. Engl.*, 2015. 54(11): pp. 3368–3386.

Frost & Sullivan Market Insight, 2011. Membrane Filtration in Southeast Asia Water and Wastewater Treatment, http://www.frost.com/prod/servlet/market-insight-print.pag?docid=265741364

Fu, F.J., S. Zhang, S.P. Sun, K.Y. Wang, and T.S. Chung, "POSS-containing delamination-free dual-layer hollow fiber membranes for forward osmosis and osmotic power generation." *J. Membr. Sci.*, 2013. 443, 144–155.

Fu, F.J., S.P. Sun, S. Zhang, and T.S. Chung, "Pressure retarded osmosis dual-layer hollow fiber membranes developed by co-casting method and ammonium persulfate (APS) treatment." *J. Membr. Sci.*, 2014. 469, 488–498.

Gao, P., Z. Liu, M. Tai, D.D. Sun, and W. Ng, "Multifunctional graphene oxide–TiO2 microsphere hierarchical membrane for clean water production." *Appl. Catal. B: Environ.*, 2013. 138–139(0): pp. 17–25.

Gao, J., S.P. Sun, W.P. Zhu, and T.S. Chung, "Polyethyleneimine (PEI) cross-linked P84 nanofiltration (NF) hollow fiber membranes for Pb^{2+} removal." *J. Membr. Sci.*, 2014. 452, 300–310.

Gao, J., S.P. Sun, W.P. Zhu, and T.S. Chung, "Chelating polymer modified P84 nanofiltration (NF) hollow fiber membranes for high efficient heavy metal removal." *Water Res.*, 2014. 63, 252–261.

Geise, G.M., B.D. Freeman, and D.R. Paul, "Characterization of a sulfonated pentablock copolymer for desalination applications." *Polymer,* 2010. 51(24): pp. 5815–5822.

Gopal, R., S. Kaur, Z.W. Ma, C. Chan, S. Ramakrishna, and T. Matsuura, "Electrospun nanofibrous filtration membrane." *J. Membr. Sci.,* 2006. 281(1–2), 581–586.

Gunawan, P., C. Guan, X. Song, Q. Zhang, S.S.J. Leong, C. Tang, Y. Chen, M. Chan-Park, B., M.W. Chang, K. Wang, and R. Xu, "Hollow fiber membrane decorated with Ag/MWNTs: Toward effective water disinfection and biofouling control." *ACS Nano,* 2011. 5(12): pp. 10033–10040.

Han, G., S. Zhang, X. Li, N. Widjojo, and T.S. Chung, "Thin film composite forward osmosis membranes based on polydopamine modified polysulfone substrates with enhancements in both water flux and salt rejection." *Chem. Eng. Sci.,* 2012. 80, 219–231.

Han, G., P. Wang, and T.S. Chung, "Highly robust thin-film composite pressure retarded osmosis (PRO) hollow fiber membranes with high power densities for renewable salinity-gradient energy generation." *Environ. Sci. Technol.,* 2013a. 47, 8070–8077.

Han, G., S. Zhang, X. Li, and T.S. Chung, "High performance thin film composite pressure retarded osmosis (PRO) membranes for renewable salinity-gradient energy generation." *J. Membr. Sci.,* 2013b. 440, 108–121.

Hyflux, 2008. Kristal® Polymeric Hollow Fibre Ultrafiltration Membrane, http://www.hydrasyst.com/innovation/hyflux-kristal.

Inge GmbH, 2013. Multibore® membranes. http://www.inge.basf.com/ev/internet/inge/en/content/inge/Produkte/Multibore_Membran

International Finance Corporation, 2007. "Environmental, health, and safety guidelines-Textile manufacturing." World Bank Group, Washington D.C.

Kaufman, Y., A. Berman, and V. Freger, "Supported lipid bilayer membranes for water purification by reverse osmosis." *Langmuir,* 2010. 26(10): pp. 7388–7395.

Kaur, S., D. Rana, T. Matsuura, S. Sundarrajan, and S. Ramakrishna, "Preparation and characterization of surface modified electrospun membranes for higher filtration flux." *J. Membr. Sci.,* 2012. 390–391(0): pp. 235–242.

Keppel, 2010. MEMSTILL® — Membrane distillation. http://www.keppelseghers.com/en/content.aspx?sid=3023

Khayet, M., and T. Matsuura, "Application of surface modifying macromolecules for the preparation of membranes for membrane distillation." *Desalination,* 2003. 158, 51–56.

Lawson, K.W., and D.R. Lloyd, "Membrane distillation." *J. Membr. Sci.,* 1997. 124, 1–25.

Lee, K.L., R.W. Baker, and H.K. Lonsdale, "Membrane for power generation by pressure retarded osmosis." *J. Membr. Sci.,* 1981. 8, 141–171.

Li, X., K.Y. Wang, B. Helmer, and T.S. Chung, "Thin-film composite membranes and formation mechanism of thin-film layers on hydrophilic cellulose acetate propionate substrates for forward osmosis processes." *Ind. Eng. Chem. Res.,* 2012a. 51, 10039–10050.

Li, X., and T.S. Chung, "Effects of free volume in thin-film composite membranes on osmotic power generation." *AIChE J,* 2013. 59, 4749–4761.

Li, X., T. Cai, and T.S. Chung, "Anti-fouling behavior of hyperbranched polyglycerol-grafted poly(ether sulfone) hollow fiber membranes for osmotic power generation." *Environ. Sci. Technol.,* 2014. 48, 9898–9907.

Li, X., R. Wang, C.Y. Tang, A. Vararattanavech, Y. Zhao, J. Torres, and A.G. Fane, "Preparation of supported lipid membranes for aquaporin Z incorporation." *Colloids Surf. B Biointerfaces*, 2012b. 94, 333–340.

Liao, Y., R. Wang, M. Tian, C. Qiu, and A.G. Fane, "Fabrication of polyvinylidene fluoride (PVDF) nanofiber membranes by electro-spinning for direct contact membrane distillation." *J. Membr. Sci.*, 2013a. 425–426, 30–39.

Liao, Y., R. Wang, and A.G. Fane, "Engineering superhydrophobic surface on poly (vinylidene fluoride) nanofiber membranes for direct contact membrane distillation." *J. Membr. Sci.*, 2013b. 44, 74–84.

Liu, C., "2.5 — Advances in Membrane Technologies for Drinking Water Purification." *Comprehensive Water Quality and Purification*. Waltham, Elsevier, 2014.

Loeb, S., and S. Sourirajan, "High flow porous membranes for separation of water from saline solutions." US patent 3133132, 1964.

Loeb, S., F.V. Hessen, and D. Shahaf, "Production of energy from concentrated brines by pressure retarded osmosis. II. Experimental results and projected energy costs." *J. Membr. Sci.*, 1976. 1, 249–269.

Loh, C. H., and R. Wang, "Fabrication of PVDF hollow fiber membranes: Effects of low-concentration Pluronic and spinning conditions." *J. Membr. Sci.*, 2014. 466, 130–141.

Luo, L., G. Han, T.S. Chung, M. Weber, C. Staudt, and C. Maletzko, "Oil/water separation via ultrafiltration by novel triangle-shape tri-bore hollow fiber membranes from sulfonated polyphenylenesulfone." *J. Membr. Sci.*, 2015. 476, 162–170.

Luo, L., P. Wang, S. Zhang, G. Han, and T.S. Chung, "Novel thin-film composite tri-bore hollow fiber membrane fabrication for forward osmosis." *J. Membr. Sci.*, 2014. 461, 28–38.

Nasreen, S.A.A.N., S. Sundarrajan, S.A. Syed Nizar, R. Balamurugan, and S. Ramakrishna, "In situ polymerization of PVDF-HEMA polymers: electrospun membranes with improved flux and antifouling properties for water filtration." *Polym. J.*, 2014. 46, 167–174.

Ong, R.C. and T.S. Chung, "Fabrication and positron annihilation spectroscopy (PAS) characterization of cellulose triacetate membranes for forward osmosis." *J. Membr. Sci.*, 2012a. 394–395, 230–240.

Ong, R.C., T.S. Chung, B.J. Helmer, and J.S. de Wit, "Novel cellulose esters for forward osmosis membranes." *Ind. Eng. Chem. Res.*, 2012b. 51, 16135–16145.

Ong, Y.K., F.Y. Li, S.P. Sun, B.W. Zhao, C.Z. Liang, and T.S. Chung, "Nanofiltration hollow fiber membranes for textile wastewater treatment: Lab-scale and pilot-scale studies." *Chem. Eng. Sci.*, 2014a. 114, 51–57.

Ong, Y.K. and T.S. Chung, "Mitigating the hydraulic compression of nanofiltration hollow fiber membranes through a single-step direct spinning technique." *Environ. Sci. Technol.*, 2014b. 48, 13933–13940.

Peng, N., M.M. Teoh, T.S. Chung, and L.L. Koo, "Novel rectangular membranes with multiple hollow holes for ultrafiltration." *J. Membr. Sci.*, 2011. 372(1–2): 20–28.

Prattle, R.E. "Production of electric power by mixing fresh and salt water in the hydroelectric pile." *Nature*, 1954. 174, 660–660.

PUB, 2010. Operation and Evaluation of Memstill Pilot Plant, http://www.pub.gov.sg/research/Key_Projects/Pages/Membrane3.aspx?Print2=yes.

PUB, 2011. Innovation in Water Singapore, Vol 1.

PUB, 2013. Innovation in Water Singapore, Vol 5.

Qi, S., C.Q. Qiu, Zhao, Y., and C.Y. Tang, "Double-skinned forward osmosis membranes based on layer-by-layer assembly — FO performance and fouling behavior." *J. Membr. Sci.*, 2012. 405–406, 20–29.

Qin, J., and T.S. Chung, "Effect of dope flow rate on the morphology, separation performance, thermal and mechanical properties of ultrafiltration hollow fibre membranes." *J. Membr. Sci.*, 1999. 157(1): pp. 35–51.

Qin, J.J., J. Gu, and T.S. Chung, "Effect of wet and dry-jet wet spinning on the shear-induced orientation during the formation of ultrafiltration hollow fiber membranes." *J. Membr. Sci.*, 2001. 182(1): pp. 57–75.

Qiu, C., L., Setiawan, R., Wang, C.Y. Tang, and A.G. Fane, "High performance flat sheet forward osmosis membrane with an NF-like selective layer on a woven fabric embedded substrate." *Desalination*, 2012. 287, 266–270.

Qu, X.Y., Zhang, L., D.S. Tang, Z.J. Zhou, and H.L. Chen, "Recovery of petroleum ether from solanesol extracting solution through vacuum hydrophilic membrane distillation." *Ind. Engin. Chem. Res.*, 2008. 47, 9544–9551.

Rigaud, J.L., B. Pitard, and D. Levy, "Reconstitution of membrane-proteins into liposomes-application to energy transducing membrane-proteins." *BBA-Bioenergetics,* 1995. 1231, 223–246.

Setiawan, L., R. Wang, K. Li, and A.G. Fane, "Fabrication of novel poly(amide–imide) forward osmosis hollow fiber membranes with a positively charged nanofiltration-like selective layer." *J. Membr. Sci.*, 2011. 369, 196–205.

Setiawan, L., R. Wang, K. Li, and A.G. Fane, "Fabrication and characterization of forward osmosis hollow fiber membranes with antifouling NF-like selective layer." *J. Membr. Sci.*, 2012. 394–395, 80–88.

Skilhagen, S.E. "Osmotic power — a new, renewable energy source." *Desalin. Water Treat.*, 2010. 15, 271–278.

Song, X., Z. Liu, and D.D. Sun, "Nano gives the answer: breaking the bottleneck of internal concentration polarization with a nanofiber composite forward osmosis membrane for a high water production rate." *Adv. Mat.*, 2011. 23, 3256–3260.

Song, X., Z. Liu, and D.D. Sun, "Energy recovery from concentrated seawater brine by thin-film nanofiber composite pressure retarded osmosis membranes with high power density." *Energy Environ. Sci.*, 2013. 6, 1199–1210.

Su, J., T.S. Chung, B.J. Helmer, and J.S. de Wit, "Enhanced double-skinned FO membranes with inner dense layer for wastewater treatment and macromolecule recycle using Sucrose as draw solute." *J. Membr. Sci.*, 2012. 396, 92–100.

Su, J., R.C. Ong, P. Wang, and T.S. Chung, "Advanced FO membranes from newly synthesized CAP polymer for wastewater reclamation through an integrated FO-MD hybrid system." *AIChE J.*, 2013. 59, 1245–1254.

Sukitpaneenit, P., and T.S. Chung, "High performance thin-film composite forward osmosis hollow fiber membranes with macrovoid-free and highly porous structure for sustainable water production." *Environ. Sci. Technol.*, 2012. 46, 7358–7365.

Sun, S.P., T.A. Hatton, and T.S. Chung, "Hyperbranched polyethyleneimine induced cross-linking of polyamide-imide nanofiltration hollow fiber membranes for effective removal of ciprofloxacin." *Environ. Sci. Technol.*, 2011. 45, 4003–4009.

Sun, S.P., T.A. Hatton, S.Y. Chang, and T.S. Chung, "Novel thin-film composite nanofiltration hollow fiber membranes with double repulsion for effective removal of emerging organic matters from water." *J. Membr. Sci.*, 2012. 401–402, 152–162.

Sun, S.P., and T.S. Chung, "Outer-selective pressure-retarded osmosis hollow fiber membranes from vacuum-assisted interfacial polymerization for osmotic power generation." *Environ. Sci. Technol.*, 2013. 47, 13167–13174.

Tang, Y.P., J.X. Chan, T.S. Chung, M. Weber, C. Staudt, and C. Maletzko, "Simultaneously covalent and ionic bridging towards antifouling of GO-imbedded nanocomposite hollow fiber membranes." *J. Mater. Chem. A*, 2015. 3(19): pp. 10573–10584.

Teoh, M.M., and T.S. Chung, "Membrane distillation with hydrophobic macrovoid-free PVDF-PTFE hollow fiber membranes." *Sep. Purif. Technol.*, 2009. 66, 229–236.

Teoh, M.M., N. Peng, T.S. Chung, and L.L. Koo, "Development of novel multichannel rectangular membranes with grooved outer selective surface for membrane distillation." *Ind. Engin. Chem. Res.*, 2011. 50, 14046–14054.

Thong, Z., G. Han, Y. Cui, J. Gao, T.S. Chung, S.Y. Chan, and S. Wei, "Novel nanofiltration membranes consisting of a sulfonated pentablock copolymer rejection layer for heavy metal removal." *Environ. Sci. Technol.*, 2014. 48, 13880–13887.

Wachinski, A.M. "Membrane processes for water reuse." The McGraw-Hill Companies, New York, 2013.

Wan, C.F., and T.S. Chung, "Osmotic power generation by pressure retarded osmosis using seawater brine as the draw solution and wastewater retentate as the feed." *J. Membr. Sci.*, 2015. 479,148–158.

Wang, H., T.S. Chung, Y.W. Tong, K. Jeyaseelan, A. Armugam, Z. Chen, M. Hong, and W. Meier, "Highly permeable and selective pore-spanning biomimetic membrane embedded with Aquaporin Z." *Small*, 2012. 8(8): pp. 1185–1190.

Wang, H., T.S. Chung, Y.W. Tong, K. Jeyaseelan, A. Armugam, H.H.P. Duong, F. Fu, H. Seah, J. Yang, and M. Hong, "Mechanically robust and highly permeable AquaporinZ biomimetic membranes." *J. Membr. Sci.*, 2013. 434, 130–136.

Wang, K.Y., T. Matsuura, T.S. Chung, and W.F. Guo, "The effects of flow angle and shear rate within the spinneret on the separation performance of poly(ethersulfone) (PES) ultrafiltration hollow fiber membranes." *J. Membr. Sci.*, 2004. 240(1–2), 67–79.

Wang, K.Y. and T.S. Chung, "Fabrication of polybenzimidazole (PBI) nanofiltration hollow fiber membranes for removal of chromate." *J. Membr. Sci.*, 2006. 281, 307–315.

Wang, K.Y., J.J. Qin, and T.S. Chung, "Polybenzimidazole (PBI) nanofiltration hollow fiber membranes applied in forward osmosis process," *J. Membr. Sci.*, 2007. 300, 6–12.

Wang, K.Y., T.S. Chung, and M. Gryta, "Modified single layer PVDF hollow fiber membrane for desalination through membrane distillation process." *Chem. Eng. Sci.*, 2008. 63, 2587.

Wang, K.Y., S.W. Foo, and T.S. Chung, "Mixed matrix PVDF hollow fiber membranes with nanoscale pores for desalination through direct contact membrane distillation." *Ind. Eng. Chem. Res.*, 2009a. 48, 4474.

Wang, K.Y., Q. Yang, T.S. Chung, and R. Rajagolapan, "Enhanced forward osmosis from chemically modified polybenzimidazole (PBI) nanofiltration hollow fiber membranes with thin wall." *Chem. Eng. Sci.*, 2009b. 64, 1577–1584.

Wang, K.Y., R.C. Ong, and T.S. Chung, "Double-skinned forward osmosis membranes for reducing internal concentration polarization within the porous sublayer." *Ind. Eng. Chem. Res.*, 2010. 49, 4824–4831.

Wang, P., M.M. Teoh, and, T.S. Chung, "Morphological architecture of dual-layer hollow fiber for membrane distillation with higher desalination performance." *Water Res.*, 2011. 45, 5489–5500.

Wang, P. and T.S. Chung, "Exploring the spinning and operation of multi-bore hollow fiber membranes for vacuum membrane distillation." *AIChE J.*, 2014a. l40, 1078–1090.

Wang, P., Luo, L., and T.S. Chung, "Tri-bore ultra-filtration hollow fiber membranes with a novel triangle-shape outer geometry." *J. Membr. Sci.*, 2014b. 452(0): pp. 212–218.

Wang, P. and T.S. Chung, "Recent advances in membrane distillation processes: Membrane development, configuration design and application exploring." *J. Membr. Sci.*, 2015. 474, 39–56.

Wang, R., L. Shi, C.Y. Tang, S. Chou, C. Qiu, and A.G. Fane, "Characterization of novel forward osmosis hollow fiber membranes." *J. Membr. Sci.*, 2010. 355, 158–167.

Wei, X., R. Wang, Z. Li, and A.G. Fane, "Development of a novel electrophoresis-UV grafting technique to modify PES UF membranes used for NOM removal." *J. Membr. Sci.*, 2006. 273(1–2), 47–57.

Widjojo, N., T.S. Chung, M. Weber, C. Maletzko, and V. Warzelhan, "A sulfonated polyphenylenesulfone (sPPSU) as the supporting substrate in thin film composite (TFC) membranes with enhanced performance for forward osmosis (FO)." *Chem. Eng. J.*, 2013. 220, 15–23.

Xing, D.Y., S.Y. Chan, and T.S. Chung, "The ionic liquid {EMIN}OAc as a solvent to fabricate stable polybenzimidazole membranes for organic solvent nanofiltration." *Green Chem.*, 2014. 16, 1383–1392.

Xu, L.Q., J.C. Chen, R. Wang, K.G. Neoh, E.T. Kang, and G.D. Fu, "A poly(vinylidene fluoride)-graft-poly(dopamine acrylamide) copolymer for surface functionalizable membranes." *RSC Advances,* 2013. 3(47): pp. 25204–25214.

Yamagishi, H., J.V. Crivello, and G. Belfort, "Development of a novel photochemical technique for modifying polu(arysulfone) ultrafiltration membranes." *J. Membr. Sci.*, 1995. 105, 237–247.

Yang, Q., K.Y. Wang, and T.S. Chung, "Dual-layer hollow fibers with enhanced flux as novel forward osmosis membranes for water production." *Environ. Sci. Technol.*, 2009. 43, 2800–2805.

Yip, N.Y., A. Tiraferri, W.A. Phillip, J.D. Schiffman, and M. Elimelech, "High performance thin-film composite forward osmosis membrane." *Environ. Sci. Technol.*, 2010. 44, 3812–3818.

Zhang, S., K.Y. Wang, T.S. Chung, H. Chen, Y.C. Jean, and G. Amy, "Well-constructed cellulose acetate membranes for forward osmosis: Minimized internal concentration polarization with an ultra-thin selective layer." *J. Membr. Sci.*, 2010. 360, 522–535.

Zhang, S., K.Y. Wang, T.S. Chung, Y.C. Jean, and H.M. Chen, "Molecular design of the cellulose ester-based forward osmosis membranes for desalination." *Chem. Eng. Sci.*, 2011. 66, 2008–2018.

Zhang, S. and T.S. Chung, "Minimizing the instant and accumulative effects of salt permeability to sustain ultrahigh osmotic power density." *Environ. Sci. Technol.*, 2013. 47, 10085–10092.

Zhang, S., F.J. Fu, and T.S. Chung, "Substrate modifications and alcohol treatment on thin film composite membranes for osmotic power." *Chem. Eng. Sci.*, 2013. 87, 40–50.

Zhang, S., M.H. Peh, Z.W. Thong, and T.S. Chung, "Thin film interfacial cross-linking approach to fabricate a chitosan rejecting layer over poly(ether sulfone) support for heavy metal removal." *Ind. Eng. Chem. Res.*, 2014. 54, 472–479.

Zhao, Y.H., K.H. Wee, and R. Bai, "Highly hydrophilic and low-protein-fouling polypropylene membrane prepared by surface modification with sulfobetaine-based zwitterionic polymer through a combined surface polymerization method." *J. Membr. Sci.*, 2010. 362(1–2): 326–333.

Zhong, P.S., T.S. Chung, K. Jeyaseelan, and A. Armugam, "Aquaporin-embedded biomimetic membranes for nanofiltration." *J. Membr. Sci.*, 2012a. 407–408, 27–33.

Zhong, P.S., N. Widjojo, T.S. Chung, M. Weber, and C. Maletzko, "Positively charged nanofiltration (NF) membranes via UV grafting on sulfonated poly-phenylenesulfone (sPPSU) for effective removal of textile dyes from wastewater." *J. Membr. Sci.*, 2012b. 417–418, 52–60.

Zhong, P.S., X.Z. Fu, T.S. Chung, M. Weber, and C. Maletzko, "Development of Thin-Film Composite forward Osmosis Hollow Fiber Membranes Using Direct Sulfonated Polyphenylenesulfone (sPPSU) as Membrane Substrates." *Environ. Sci. Technol.*, 2013. 47, 7430–7436.

Zhu, W.P., J. Gao, S.P. Sun, S. Zhang, and T.S. Chung, "Poly(amidoamine) dendrimer (PAMAM) grafted on thin film composite (TFC) nanofiltration (NF) hollow fiber membranes for heavy metal removal." *J. Membr. Sci.*, 2015. 487, 117–126.

Zhu, X., H.E. Loo, and R. Bai, "A novel membrane showing both hydrophilic and oleophobic surface properties and its non-fouling performances for potential water treatment applications." *J. Membr. Sci.*, 2013. 436(0): pp. 47–56.

4. Nanostructured Catalytic and Adsorbent Materials for Water Remediation

Zhong Chen* and Teik Thye Lim[†]

1. Introduction and Overview

This chapter is dedicated as an overview of the research and development in Singapore on the catalysts and adsorbents for water remediation. Catalytic water treatment belongs to the broad category of advanced oxidation processes (AOPs). AOPs make use of a series of chemical reactions to remove organic and inorganic pollutants in water by oxidation. Two types of catalytic water treatment will be covered in this chapter. Photocatalysis works through absorption of photons on a semiconductor material with a suitable band gap. The absorbed photons excite the valence electrons to the conduction band of the semiconductor and this generates electron-hole pairs. The separated electron-hole pairs will then produce radicals for water treatment reactions (reduction and oxidation). The Fenton process makes use of H_2O_2 and an iron catalyst for the oxidation of contaminants. During the Fenton process, Fe^{2+} and Fe^{3+} are shuttled (by reduction and oxidation) by H_2O_2 leaving behind radicals for chemical treatment reactions. On the other hand, unlike the catalytic treatments just described, adsorption, as an integral part of water remediation technologies, is a physical process whereby the pollutant ions or molecules are concentrated for subsequent disposal.

---

*School of Materials Science and Engineering, Nanyang Technological University, ASZChen@ntu.edu.sg.
[†]School of Civil and Environmental Engineering, Nanyang Technological University, cttlim@ntu.edu.sg.

As the book title suggests, we will limit the review on the research effort made in Singapore on materials for water remediation. Most of the references are about the research conducted in Singapore in the past 10 years. References to research work/report done elsewhere are only to assist in relating and giving an account of the historic development and for the completeness of the information and discussion. The works on photocatalysts will be covered in the next section. Photocatalytic degradation of pollutants is by far the most researched topic for catalytic water treatment worldwide, and this is no exception for the research conducted in Singapore. An increasing number of research papers have been published in the last 10 years, especially more so since 2010. Therefore from this reservoir of information we can only selectively highlight some key results based on our limited knowledge and (probably subjective) judgement. In Section 3, chemical catalysts such as the Fenton process will be reviewed. Section 4 focuses on adsorbent materials and adsorbent-photocatalyst coupling. Finally, we conclude the chapter with our views on future perspectives in this important and rapidly growing field.

2 Photocatalysts

2.1 TiO_2 and other semiconductor compounds

TiO_2 is by far the most investigated material for photocatalytic applications. This is due to its low cost, favorable band gap positions, non-toxicity and inertness in a wide range of pH conditions. TiO_2 may exist in over 10 different crystalline polymorph structures, but there are only 4 TiO_2 polymorphs that have been synthesized and studied for photocatalytic reactions, namely rutile, anatase, brookite, and TiO_2(B) (Fig. 1). In a TiO_2 crystal, a titanium cation is six-fold coordinated to the surrounding oxygen anions, forming distorted TiO_6 octahedra. The TiO_6 octahedral building blocks are joined by edge-sharing in different spatial arrangements leading to different polymorph structures. Different crystal structures lead to different materials properties, which makes this family of materials more interesting for research and applications.

Anatase and rutile are the most studied polymorphs of TiO_2 for photocatalytic applications. Both have an indirect band gap, which is 3.0 eV for rutile and 3.2 eV for anatase, corresponding to ultra-violet (UV) light absorption. It is generally believed that anatase is the most active phase for photocatalysis applications, due to its better electronic and surface chemistry properties. That is the reason for most photocatalytic research being conducted using anatase TiO_2 or P25, a commercial product in nanoparticle form consisting of mixed anatase and rutile

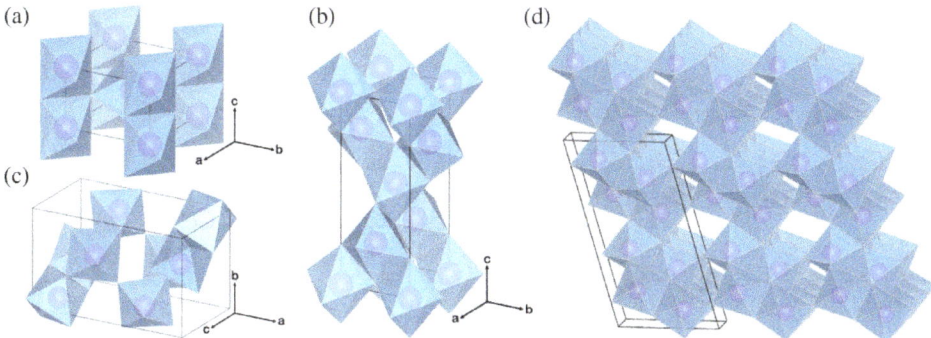

Fig. 1. TiO$_2$ polymorph structures: (a) Rutile; (b) Anatase; (c) Brookite; and (d) TiO$_2$(B). Purple spheres represent a Ti atom, and the blue octahedra represent TiO$_6$ blocks. Oxygen atoms (not shown) are at the corners of the octahedra.

phases. TiO$_2$ has been extensively studied for its degradation ability of various organic compounds,[1–35] and the detoxification of metal ions by either oxidation or the reduction of the ions from a toxic valence state to a less or non-toxic state.[36–39] It is worth mentioning that a great deal of effort has been made in the past decades to increase the surface area of TiO$_2$ by nanostructure formation,[19,29,40] to couple TiO$_2$ with another metal or semiconductor for prolonging the charge lifetime,[20,22,25–28,31,34,41] and to prepare TiO$_2$-based membranes for multi-functional water treatment.[20,21,33] Meanwhile, an investigation on the photocatalytic mechanisms was reported by J.G. Highfield et al.[42] using in-situ Diffuse Reflectance Infrared Fourier Transform Spectroscopy (DRIFTS) coupled with an on-line Mass Spectrometer (MS). D.G. Gong et al.[43] studied the intermediates during photocatalytic oxidation of ethanol over pristine and platinized P25 TiO$_2$ by monitoring the solid-liquid interface using Attenuated Total Reflectance (ATR) Fourier Transform Infrared (FTIR) spectroscopy.

TiO$_2$(B) was reported to have a similar band gap to anatase at 3.2 eV, and a theoretical calculation indicates that the conduction band edge position is more cathodic than anatase[44]. In theory, this would make it more active than anatase for photocatalytic reductive reactions and less active than anatase for oxidation. Experimental work is rare for TiO$_2$(B) when being used alone as a photocatalyst, probably because of its relatively weaker oxidation power when compared with anatase. However, some literatures reported excellent activity when TiO$_2$(B) is coupled with anatase to form a heterojunction structure.[44–47] The charge transfer between the coupled semiconductors enabled a prolonged charge life, leading to improved activity.

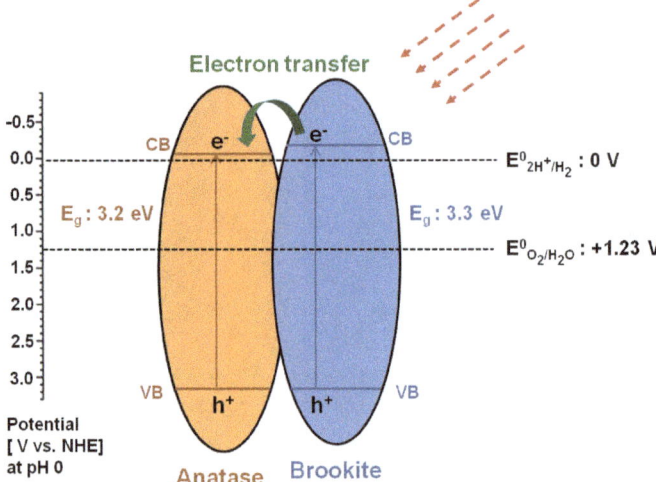

Fig. 2. Schematic of the transfer of a photo-generated electron from the conduction band of brookite to the conduction band of anatase under UV-Vis irradiation. The positions of as-synthesized anatase and brookite TiO_2 band edges relative to the redox potential of water vs NHE are determined from a Mott-Schottky analysis.[50]

Brookite TiO_2 is another relatively rarely studied material, and it was reported that brookite nanocrystals have higher photocatalytic activities than rutile and anatase.[48,49] Z. Chen's group[50] at NTU synthesized highly crystalline single phase brookite and coupled anatase/brookite TiO_2 nanoparticles via a simple one-pot hydrothermal process. The band gap of brookite TiO_2 was found to be 3.3 eV, and its band gap potential is ~0.1 eV more negative than anatase (Fig. 2).[50] Using hydrogen yield as a measure of the photocatalytic activity, the two-phase anatase/brookite TiO_2 shows the best activity when compared with single phase brookite or anatase. They also proved that indeed the bookite nanoparticle is more active than anatase. As explained by Fig. 2, photoexcited electron transfer from the brookite to the anatase phase is the reason leading to an enhancement of the photocatalytic activity. In comparison with the highly active two-phase commercial benchmark P25, the two-phase anatase/brookite TiO_2 is 2.2 times more active per unit area of the photocatalyst's surface.[50]

While TiO_2 has been most extensively studied for photocatalytic applications, efforts to explore other semiconductor compounds have intensified in recent years. The incentive to study alternative photocatalysts comes from the interest to discover more efficient uses of solar energy (e.g. using pristine visible light active photocatalyst), more active catalyst materials, and lower production costs for practical applications. The materials being investigated include CdS,[5,31,51–67] ZnO,[40,41,68–85] carbon nitride,[86–92] WO_3,[93–95] $BiVO_4$,[96–100] $NaTaO_3$,[101–106] and others.[107–130] The same incentive has also been the driving force to improve the photocatalytic activity of TiO_2. Researchers have been working on enhancing the photo-activity of TiO_2 by doping (to be discussed next), exposing more active crystal facets,[23,106,131–137] and forming heterojunction structures to delay the intrinsically fast charge recombination.[22,50,136,138–143]

2.2 Metal/non-metal doped TiO_2 for visible-light photocatalysis

Due to the wide band gap of TiO_2 and other stable wide band-gap semiconductor oxides, several strategies have been applied to enable the visible light activity of these photocatalysts. These include doping with metal or non-metal element(s), loading of a visible light sensitizer (usually another semiconductor material with a narrow band gap), and using metal particles to create the so-called plasmonic effect (to be discussed in the next sub-section).

T.T. Lim's group at NTU has produced several modified TiO_2 which were doped/co-doped with nonmetals and metals to produce visible-light photoresponsive TiO_2.[144–149] The doped TiO_2 were evaluated for visible-light photocatalytic degradation of various recalcitrant pollutants including antibiotics, pharmaceuticals, plasticizers, pesticides and dyes. The toxicity evolution of the solution during the photocatalytic treatment was investigated to evaluate the combined toxicity of the intermediate and end products of the degradation as compared to that of the parent compounds. In some of their studies, simulated wastewater samples were tested which contained humic acids and electrolyte species commonly found in surface water, groundwater and wastewater.[145,146,148,150] They also attempted to identify the production of various reactive oxygen species (ROS) through the use of various selective scavengers.[144,151]

C-N co-doped TiO_2 (C-N-TiO_2) was synthesized by a solvothermal method.[145] The effects of the source of dopants, dopant concentration, solvothermal treatment condition, and the temperature of calcination were evaluated. The C-N-TiO_2 showed a much higher removal efficiency of bisphenol A than that of C-TiO_2, N-TiO_2 and P25 under white LED irradiation. Under the simulated solar light and UV irradiation, bisphenol A was almost completely mineralized by the C-N-TiO_2 within 5 h. After 5 h of reaction, the bisphenol A solution became almost non-toxic. The C-N-TiO_2 was still able to remove a small fraction of bisphenol A even under yellow LED irradiation. With the use of various scavengers, findings suggested that $O_2^{\cdot-}$ produced from the photoreduction of dissolved oxygen could be the predominant oxidative species in the photocatalytic degradation process which takes place under visible light irradiation. The C-N-S tri-doped TiO_2 was synthesized using sol-gel method with titanium butoxide as titanium precursor and thiourea as the dopant source.[152] The photocatalyst with a thiourea : Ti molar ratio of 0.05:1 and calcined at 450 °C exhibited the highest photocatalytic activity. This could be attributed to the synergistic effects of tetracycline (TC) adsorption and band gap narrowing resulting from C-N-S tri-doping, and the well-formed anatase phase. The slightly alkaline pH condition was more favorable for the TC degradation. C-sensitized and N-doped TiO_2 (C/N-TiO_2) was also synthesized for the photocatalytic degradation of sulfanilamide under visible light LED irradiation.[146] The synthesized C/N-TiO_2 exhibited a good photochemical stability and its light absorption onset could be extended to 600 nm. The C/N-TiO_2 calcined at 300°C exhibited the highest photocatalytic activity for sulfanilamide degradation, due to the presence of an optimal content of carbonaceous species (serving as photosensitizer) and the nitrogen doping. The degradation of sulfanilamide was more efficient in the acidic condition due to the carbon photosensitizing effect. However, bicarbonate, phosphate and silica could inhibit the sulfanilamide mineralization to different degrees.

Anatase Fe-C co-doped TiO_2 (Fe-C-TiO_2) was also synthesized by a solvothermal method.[153] The synergistic effects of Fe and C co-doping resulted in the improved photocatalytic activities of Fe-C-TiO_2 for the degradation of bisphenol A and clofibric acid as compared to C-TiO_2, Fe-TiO_2 and P25 under visible-light LED and simulated solar light irradiation. The toxicities of a bisphenol A solution gradually decreases throughout its mineralization process.

An Fe-Er co-doped TiO_2 (Fe-Er-TiO_2) was synthesized using titanium isopropoxide as a titanium precursor and carbon source, as well as ferric nitrate and erbium nitrate as the dopant sources, via the solvothermal route.[148] Under visible light irradiation, the photocatalytic activity of Fe-Er-TiO_2 for the degradation of bisphenol A was higher than the pristine TiO_2, Er-TiO_2, Fe-TiO_2 and P25. The enhanced photocatalytic activity of Fe-Er-TiO_2 was believed to be due to the combined effects of photosensitizing (C-O band), narrowed band gap, enhanced e^-/h^+ separation (Ti-O-Fe linkage), and up-conversion luminescence property (Ti-O-Er linkage).

Z. Chen's group at NTU investigated Bi^{3+} doped $NaTaO_3$ for visible light photocatalytic dye degradation and hydrogen generation.[103–105] By controlling the stoichiometry of the starting chemicals, they showed that equal Bi occupancy at the Na and Ta sites leads to the visible light absorption of up to 550 nm. Correspondingly, the highest photocatalytic activity for methylene blue degradation under visible light was obtained for this condition.[105] They further carried out a first principles calculation to select potential dopants for the visible light photocatalysis of $NaTaO_3$.[102] Doping of non-magnetic cations (Ti, V, Cu, Zn, W, In, Sn, Sb, Ce, and La), anions (N, C, I), and certain co-dopant pairs (W-Ti, W-Ce, N-I, N-W, La-C, Pb-I, Cu-Sn, Cu-N) were investigated. The calculations suggest that substitutional doping of Cu at the Ta site, Cu at the Na site, and C at the O site narrows the band gap of $NaTaO_3$ from 4.0 to 2.8, 2.3, and 2.1 eV, respectively, inducing visible light absorption. Additionally, passivated co-doping of Pb-I and N-W narrows the band gap of $NaTaO_3$ to the visible region, while maintaining the band potentials at the favorable positions.

2.3 Plasmonic photocatalysts

Noble metal nanoparticles (e.g., Ag, Au, Pt) exhibit strong light absorption in the UV-vis spectral region due to the surface plasmon resonance (SPR) effect. Such an effect is dependent on the size distribution, shape and surrounding medium.[154–157] In a noble metal-decorated photocatalyst composite, the metal nanoparticles could serve as a major photoactive component, and the interface at the nanojunction of two components could efficiently separate the photogenerated electrons and holes. In recent years, a number of visible-light plasmonic photocatalysts composed of noble metal nanoparticles and metal oxide semiconductors have been developed.[79,158–182] Typical support materials are mostly semiconductors, including TiO_2,[14,163,167–172,174,176,178,179,181,183–187] ZnO,[79,163,180,188,189] Ag halide,[35,162,173,182,190,191] and other compounds.[97,112,192,193] Other materials such as carbon,[194] graphene,[195] graphene oxide[196] oxide insulators[197] and graphene-semiconductor composite[182] have also been used as support and/or charge carriers for fabricating plasmonic photocatalysts.

Several plasmonic Ag-AgBr/TiO_2 composites were synthesized by T.T. Lim and co-workers at NTU. The materials were evaluated for the photocatalytic degradation of ibuprofen, bisphenol A and Rhodamine B.[181] The Ag-AgBr particles with high uniformity were evenly distributed in the matrix of TiO_2 particles. Under yellow LED irradiation, the Ag-AgBr/TiO_2 still showed a relatively high photocatalytic activity. Among the various oxidative species (h^+, •OH, O_2^- and $^1O_2$) analyzed, O_2^- was found as the predominant species involved in the pollutant degradation process. Graphene was also introduced as a support with long-range charge carriers to further enhance the SPR effect on the photocatalytic activity of the plasmonic photocatalyst.[182]

Z.L. Dong and Z. Chen's groups at NTU successfully prepared Ag nanoparticle based plasmonic photocatalysts for visible light degradation of organic dyes.[167,170,171,173] A new strategy was designed to enable the formation of AgX (X = Cl, Br, I) nanoparticles on a surface made of Ag-titanate nanowires. The uniformity and density of the AgX NPs were controlled over the reaction time, pH value, and concentration of the HX (X = Cl, Br, I) solutions reacting with the Ag-titanate (Fig. 3). The densely distributed AgX nanoparticles with high uniformity can be obtained at acid conditions within a short time. After visible light irradiation, the assembled material consisting of Ag/AgX nanoparticles anchored onto a titanate nanowire surface exhibited very good photocatalytic activity for the decomposition of methyl orange,[170,173] methylene blue,[171] and phenol solutions.[167]

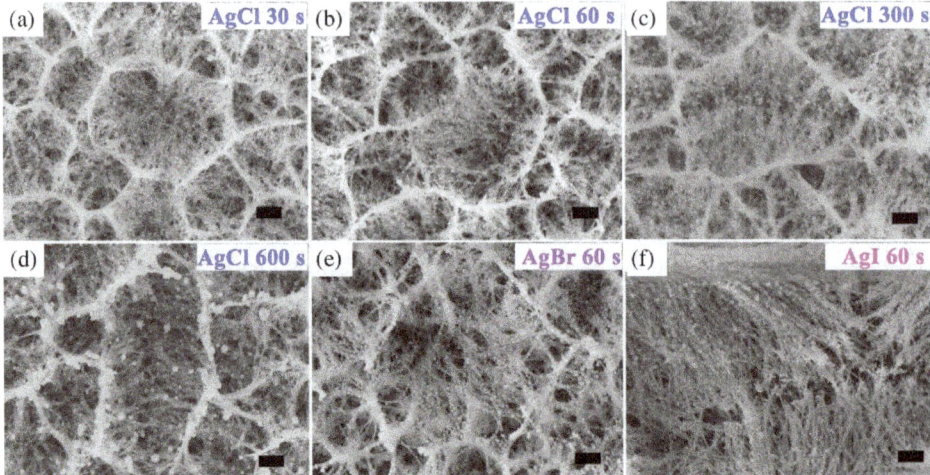

Fig. 3. Scanning electron microscopic images of AgCl on an Ag-titanite surface after reacting with a HCl solution (0.10 M) with different durations (a) 30 s; (b) 60 s; (c) 300 s; (d) 600 s. (e) and (f) are AgBr and AgI on the same titanite surface after reaction with HBr and HI solution (0.10 M) with durations of 60 s respectively. Scale bar: 1.0 μm.[280]

Similar to metal particle loading, other visible light sensitive surface groups would achieve the same effective of utilizing a broad spectrum of solar energy. Y.H. Cheng et al.[139] found that by introducing triethylamine [N(CH$_2$CH$_3$)$_3$] as the N source, solvothermally synthesized dual-phase titanate/titania showed visible light absorption beyond 800 nm. Such behavior is distinctively different from nitrogen doping, and the authors believed that the presence of surface-bond Ti-O-N chromophores could be the reason.

2.4 Photocatalytic disinfection

A wide range of pathogenic microorganisms including Gram-negative and Gram-positive bacteria, fungi, algae, protozoa and viruses, can be inactivated or killed by photocatalysis.[34,35,57,67,70,79,96,124,185,198–206] Bacteria inactivation over photo-excited TiO$_2$ can be due to initial cell membrane and wall damage, following thereafter with the destruction of intracellular components. In most studies, the reactive oxygen species are responsible for the bacteria inactivation, and •OH is suggested to be the primary oxidative species.[207,208] The reactive oxygen species can cause lipid peroxidation leading to the loss of the membranes' respiratory activity.[209] They can also result in damage to the respiratory system within the cells and the increased ion permeability in the cell membrane.[210] Furthermore, the damage of cell membranes can induce further oxidative attack of internal cellular components, eventually leading to cell death.[211] Photocatalytic destruction

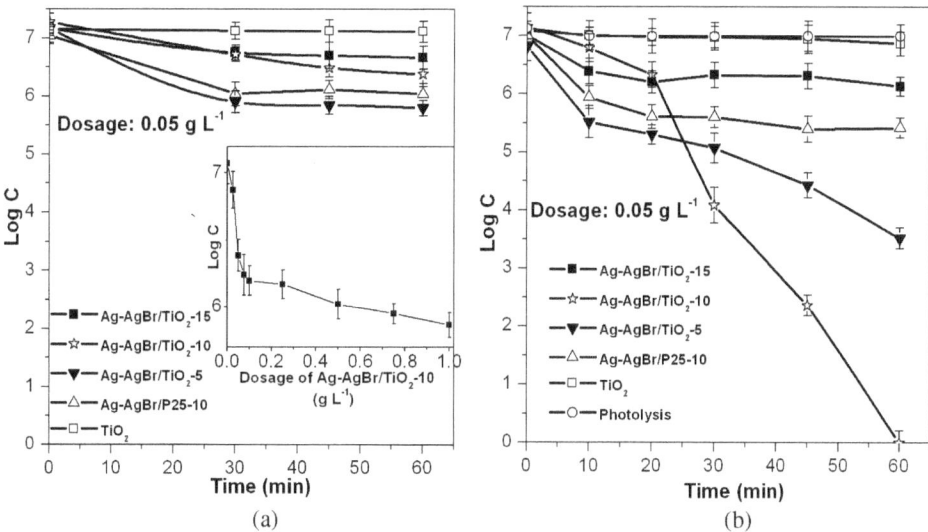

Fig. 4. E. coli inactivation by Ag-AgBr/TiO$_2$-x in the dark (a) and under white LED irradiation (b) where "x" indicates the atomic ratio of Ti/Ag. Inset of A shows the effect of Ag-AgBr/TiO$_2$_10 dosage on its antibacterial activities against E. coli in the dark.[35]

of the E. coli by the Ag-AgBr/TiO$_2$ composites was studied by Wang and Lim.[35] The findings are presented in Fig. 4. In the dark, these composites showed intrinsic antibacterial activities against Gram-negative E. coli. Under visible-light LED irradiation, both the photocatalytic disinfection activity and the intrinsic antibacterial property of Ag-AgBr/TiO$_2$ contributed to the inactivation of the E. coli, and a max of 7-log inactivation of the E. coli could be achieved within 60 min of exposure time (Fig. 4b). It was reported that even under the yellow LED irradiation, the Ag-AgBr/TiO$_2$ still showed relatively high antibacterial performance.[35] The inactivated E. coli did not show dark repair after the photocatalytic process, indicating permanent death of the pathogen.

D.D. Sun's group prepared GO-supported photocatalysts for photocatalytic water disinfection.[34,67] With GO-TiO$_2$-Ag, 7-log inactivation of the E. coli was achieved within 90 min under visible light irradiation, compared to less than 1-log inactivation even after 120 min of exposure in the dark.[34] With GO-CdS composite, nearly 90% of Gram-positive B. subtilis was killed in 10 min.[67]

2.5 Photocatalytic membrane

Photocatalytic membrane is a water remediation product featuring synergistic coupling of photocatalysis and membrane filtration. In the photocatalytic membrane reactor, mass transfer of the pollutant to the photocatalytic surface can be accelerated through forced permeation exerted by the applied transmembrane pressure

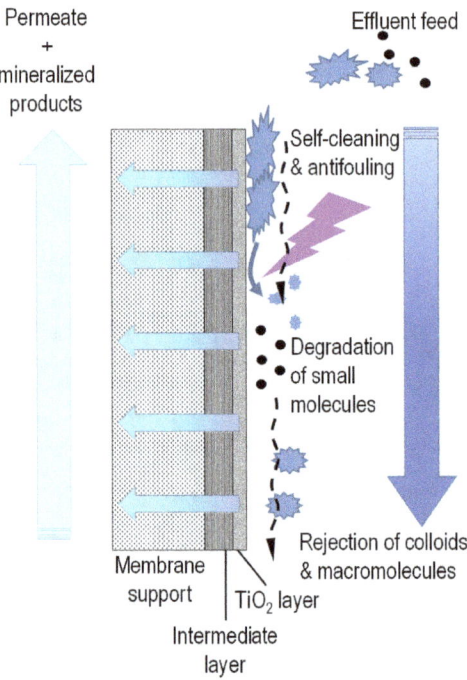

Fig. 5. Schematic illustration of synergistic coupling of photocatalytic degradation and separation processes in a photocatalytic membrane reactor with cross flow mode.

while the photocatalyst coated on the membrane surface prevents membrane fouling through a photocatalytic cleaning process. The synergistic photocatalysis-membrane filtration process is illustrated in Fig. 5.

A.G. Fane et al.[212] at NTU evaluated 10 polymer membranes for possible use in photocatalytic-membrane processes under ultraviolet irradiation with the presence of TiO_2 as the photocatalyst. Membrane stability was characterized by changes in the pure water flux, release of total organic carbon and scanning electron microscope morphology analysis. Their results revealed that polytetrafluoroethylene (PTFE), hydrophobic polyvinylidene fluoride (PVDF) and polyacrylonitrile (PAN) membranes possessed the greatest stability. However, the oxidative stability of the PAN membrane was not very good when exposed to 10 days of 200 mM H_2O_2/UV conditions.

Bai et al.[85] at NTU developed a photocatalytic polymer membrane through depositing TiO_2 nanofibers and ZnO nanorods onto a commercial cellulose membrane. Pressurized water was used to enhance adherence of the photocatalytic powders on the membrane. The resultant membrane was tested for photocatalytic oxidation of dye, water filtration and antibacterial action. Since polymeric membranes are generally susceptible to degradation during the photocatalytic treatment of water and to ensure strong adherence of photocatalyts on the coated membrane, Lim and his co-workers developed hierarchically porous photocatalytic ceramic membranes via layer-by-layer coating of different titanium sols (with different structure directing agents and concentrations) onto the macroporous alumina ceramic membrane support.[33,213] The resulting multilayer TiO_2 film possessed a porosity gradient that gave the highest water permeation and meanwhile maintained robust photocatalytic activity against membrane fouling. If loaded with Ag nanoparticles, the photocatalytic membrane became antibacterial for preventing the membrane biofouling.[214]

3 Catalysts for Advanced Oxidation Technology

3.1 Fenton and photo-fenton advanced oxidation technology

Advanced oxidation technology (AOT) presents advantages over adsorption and membrane separation technology in water remediation because AOT can assure degradation and even mineralization of various organic pollutants. Its reaction is fast due to the generation of powerful •OH with an oxidative potential of 2.8V vs NHE. For the treatment of recalcitrant pollutants, AOT is preferred over the biological treatment process. Fenton and photo-Fenton processes are among the widely applied AOTs for water remediation. While homogenous systems are still the mainstream AOTs, heterogeneous Fenton and photo-Fenton AOTs have been explored recently because they allow the recovery and reuse of the Fenton catalysts, reduction in sludge production, and expansion of the operating pH range to encompass the common pH range of water and wastewater.[215]

The conventional heterogeneous Fenton catalysts are iron oxide/hydroxide minerals as well as iron powder. Hu et al.[216] have synthesized single-crystal bismuth ferrite ($Bi_2Fe_4O_9$) for use as a Fenton and photo-Fenton catalyst. The mineral also possessed semiconductorial photocatalytic activity. The nanopad catalyst exhibited remarkable activity for visible-light photo-Fenton oxidation, dark Fenton oxidation and visible-light photocatalysis. It can be used to degrade organic pollutants in water treatment usage through multiplex catalytic oxidation processes, using sunlight in daytime while H_2O_2 should be added during the night time only. Because the bismuth ferrite also possesses the magnetic property, it can be recovered from the reactor via magnetically-enhanced gravity separation.

3.2 Sulfate radical advanced oxidation technology

Sulfate radical based-advanced oxidation technology (S-AOT) has emerged as an alternative to the AOTs based on hydroxyl radicals for water remediation. The S-AOT employs highly-reactive sulfate radicals to decompose recalcitrant organic pollutants. Sulfate radicals can be generated from peroxydisulfate ($S_2O_8^{2-}$, $E° = +2.12$

vs NHE) or peroxymonosulfate (HSO_5^-, $E°$ = +1.82 vs NHE) activation via thermal, UV, base, quinones and transition metals. The transition metal activation processes are as illustrated below:

$$Me^{n+} + S_2O_8^{2-} \rightarrow Me^{(n+1)+} + SO_4^{\bullet-} + SO_4^{2-} \quad (1)$$

$$Me^{n+} + HSO_5^- \rightarrow Me^{(n+1)+} + SO_4^{\bullet-} + OH^- \quad (2)$$

$$Me^{(n+1)+} + HSO_5^- \rightarrow Me^{n+} + SO_5^{\bullet-} + H^+. \quad (3)$$

For peroxydisulfate activation by a transition metal, it is a one-way process (Eq. (1)) without replenishing the transition metal for further reuse. On the other hand, for peroxymonosulfate activation by a transition metal, it is thermodynamically feasible for the redox cycling of the oxidized transition metal (Eqs. (2) and (3)). The $SO_4^{\bullet-}$ radical is very reactive with a relatively high standard reduction potential of +2.5–3.2 V vs NHE. On the contrary, the $SO_5^{\bullet-}$ radical is least desirable as it reacts slowly with organic compounds due to its relatively lower standard reduction potential but serves to regenerate the catalyst.

Transition metal activation of peroxymonosulfate to produce sulfate radicals for pollutant abatement can be achieved through both the homogeneous or heterogeneous catalytic systems. To avoid having to recover the catalysts after treatment, the heterogeneous system is desirable over the homogeneous reaction system. Oh et al. developed a dipicolinic acid-functionalized hematite (DPA-hematite) with a high surface area via the co-precipitation of a Fe(III)-DPA complex.[217] It was used as a catalyst to activate peroxymonosulfate for bisphenol A detoxification. The DPA-hematite can be recovered using magnet and reused by a simple washing procedure with water. Oh et al. further developed a novel 3-D $CuBi_2O_4$ consisting of self-assembled spherical nanocolumn arrays which can activate peroxydisulfate and peroxymonosulfate (Fig. 6).[218] The bifunctional catalyst was produced via a facile hydrothermal method without using any template or surfactant, and thus this was found to be a scalable clean production process.

4 Adsorbent for Removing Pollutants from Aqueous Media

4.1 Adsorbents: materials and technology

Adsorption technology is considered as one of the most appealing methods for pollutant removal because it is cost-effective, versatile, and simple to implement. While activated carbon (AC) adsorbents are cost-effective for removing most organic pollutants, they are generally not effective in removing inorganic pollutants, including

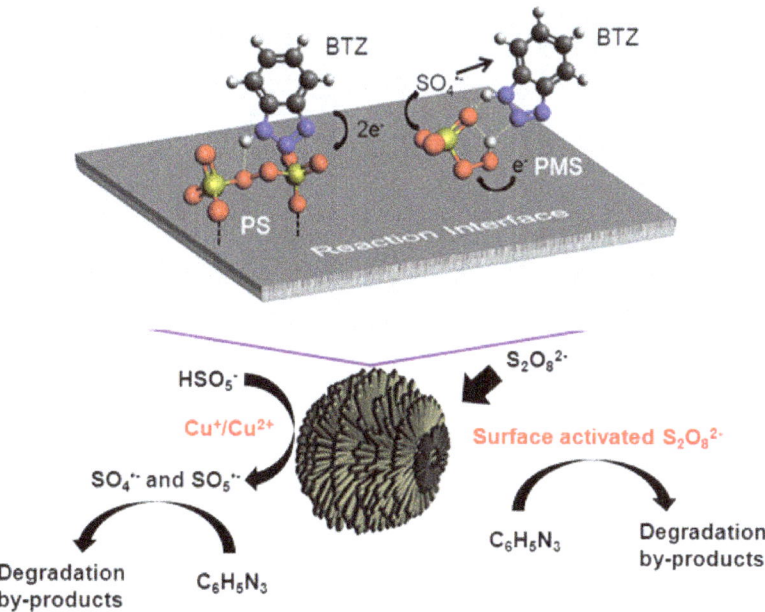

Fig. 6. Schematic illustration of peroxydisulfate (PS) and peroxymonosulfate (PMS) activations by 3-D $CuBi_2O_4$ nanocolumn arrays for 1H-benzotriazole (BTZ) degradation. [218]

toxic metals, metalloids and halides. Toxic metals belonging to cationic species can be removed through precipitation as metal hydroxides. However, the anionic forms of toxic metals, metalloids and non-metals, such as oxyanions of chromate/dichromate, arsenite/arsenate, vanadate, molybdate, selenite/selenate, etc. are highly soluble and mobile.

To remove anionic pollutants, the positively charged nanostructured materials are potentially effective adsorbents. Layered double hydroxides (LDHs) have been recently developed and investigated for effective oxyanion removal owing to theirs large surface area, high anion exchange capacity, good thermal stability, and ability to reconstruct their original layered structure during a calcination-rehydration cycle through their "memory effect".[219] Goh et al. synthesized various Mg/Al LDHs through five different synthesis routes.[220] Generally, the synthesized LDHs showed excellent adsorption of polyvalent oxyanions (> 99% removal efficiency) as compared to the commercial LDH (< 60% removal efficiency) but they were not effective for removing monovalent oxyhalides and borate. The adsorption mechanism involves both outer-sphere complexation and ligand exchange mechanism (i.e., inner-sphere complexation).[221] Arsenic could be removed in amounts of up to ~90 mg g^{-1} of the synthesized LDH.[222] The adsorbate-loaded LDHs could be easily regenerated using NaOH, $NaNO_3$ and Na_2CO_3 solutions.

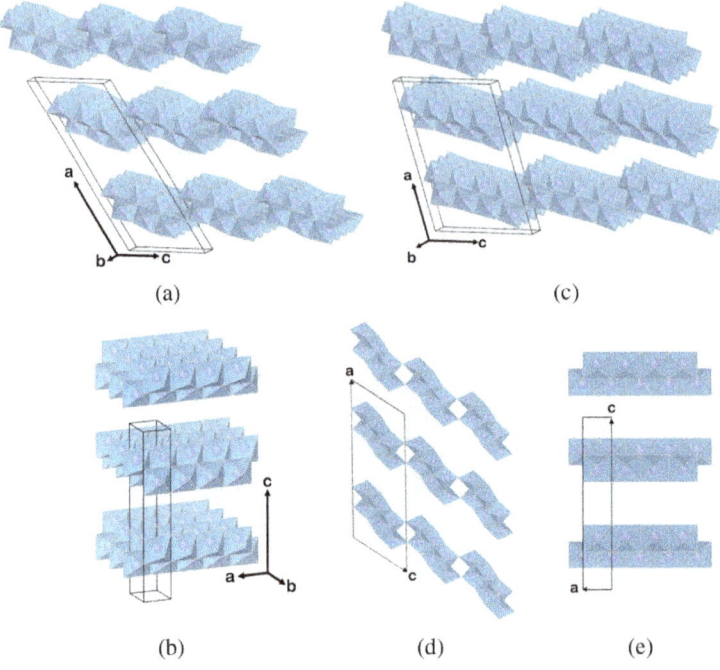

Fig. 7. Crystal structure of (a) monoclinic titanate $H_2Ti_3O_7$; (b) orthorhombic titanate $H_xTi_{2-x/4}\square_{x/4}O_4$; and (c) monoclinic titanate $H_2Ti_4O_9$. The (010) plane projection of (d) monoclinic titanate $H_2Ti_3O_7$, and (e) orthorhombic titanate. Purple spheres represent Ti atom, and the blue octahedra represent TiO_6 blocks. Oxygen atoms are located at the corner of the octahedra, other cations and molecules in the interlayers are omitted for clarity.

The open crystal structure and large interlayer spacing of titanate materials make them very suitable to accommodate many varieties of cations and neutral molecules in the interlayer spaces. Titanate crystals have two basic shapes: monoclinic and orthorhombic (Fig. 7). A large dipole moment exists in titanate materials, where the negatively charged titanate layer is balanced by positively charged interlayer cations. The unique crystal feature endows this group of materials unique physiochemical properties, for example, they possess very good cation exchange capability. And because of the negative charge on the titanate sheet, they can also effectively adsorb positively charged molecular species on the surface. These effects have been applied in wastewater remediation for the treatment of radioactive ions and organic dyes.

In the past, high surface area titanates were usually synthesized via a hydrothermal process for as long as 18 to 48 hours. Tang et al.[223] at NTU invented an ultrafast electrochemical spark discharge spallation process to produce spherulite-shaped titanate particles in the order of minutes. The synthesized particles (Fig. 8) had a very high specific surface area (> 400 m²/g) and demonstrated an extremely fast adsorption rate and large adsorption capacity for Pb^{2+} ions and methylene blue

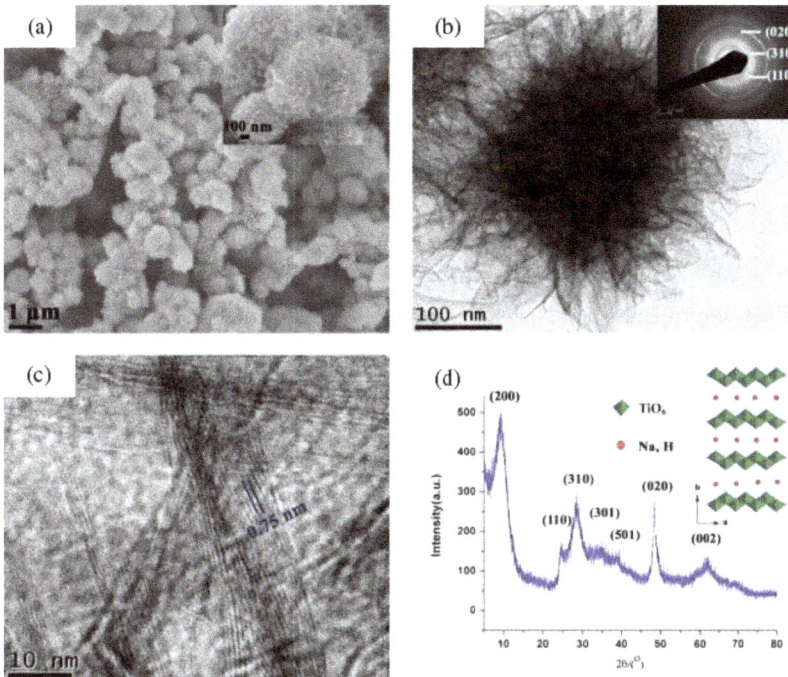

Fig. 8. (a) FESEM image, (b) TEM image, (c) high resolution TEM image, (d) XRD pattern of the fabricated titanate particles. The particles were collected after anodization of Ti foils in 10 M NaOH aqueous solution for 20 min. Inset shows the crystal structure for lepidocrocite titanates.[223]

Table 1. Comparison of adsorption capacity with different adsorbents[224]

Adsorbent sample	BET surface[a]	Removal capacity for Pb^{II}[b]
Zeolite[225]	—	0.44
Magnesium silicate[226]	355.21	0.314
Actived carbon[227,228]	612, 686	0.64, 0.56
Clay[229]	156	0.25
Polymer resin[230]	—	2.1
Sodium trititanate[224]	< 50	0.56
Titanate nanofiber[231]	—	1.36
Titanate spherulite particles[224]	406	2.42

[a] (m^2 g^{-1}); [b] (mmol g^{-1}); — not available.

in solution. Table 1 compares the adsorption capacity of the prepared spherulite particles with some available literature data. It is clear that the spherulite titanate out-performed other adsorbent materials.

In another study, Y.W.L. Lim et al.[232] synthesized TiO_2 and sodium titanate nanostructures with controllable phases and morphologies using a hydrothermal method with titanium disulfide (TiS_2) as the starting material. Sodium titanate nanobelts were synthesized under a condition with a relatively low alkaline concentration. The sodium titanate nanobelts possessed an adsorption capacity as high as 312.5 mg g^{-1} for methylene blue, and the adsorption kinetics followed a pseudo-second-order kinetic model.

J.P. Chen's group at NUS had synthesized various adsorbents for adsorption of organic and inorganic arsenic.[233–236] Using their Zr-Mn binary hydrous oxide, As(III) could be oxidized to As(V) by the MnO$_2$ component, resulting in a better As adsorption by the ZrO$_2$ component.[236] The organic arsenic could be removed using N-methylglucamine modified chitosan polymeric adsorbent and a calcium alginate encapsulated magnetic adsorbent.[235]

Fluoride contamination in drinking water is one of the major problems worldwide because excessive fluoride intake can cause the softening of bones, the ossification of tendons and ligaments, and neurological disorders. J.P. Chen's group had also developed effective adsorbents for fluoride removal.[237] Their sulfate-containing Mn-La metal composite showed the maximum adsorption capacity of 293 mg g^{-1} for fluoride. The removal mechanism involves the exchange of the sulfate ions in the composite with fluoride ions.

R. Xu's group at NTU produced several nanostructured adsorbents for the adsorption of organic pollutants such as dyes and pharmaceuticals.[238–240] Their carbon nanosphere produced from the simple hydrothermal treatment of glucose solution was enriched with –OH and –COO– functional groups and could adsorb dyes up to 680 mg g^{-1}.[241] Their Fe$_3$O$_4$-loaded LDH could form a stable colloidal suspension in water and the colloids could be separated/re-dispersed using a magnetic field.[238] A nanocomposite comprising 3-D hierarchical LDH grown on silica spheres was produced which could be monodispersed in water.[239] Lou and his co-workers produced a novel uniform urchin-like hollow sphere nanoadsorbent comprising single-phase α-FeOOH (Fig. 9) whose morphology and hollow structure could be controlled by tuning the glycerol/water volume during synthesis.[242] The FeOOH hollow sphere (Fig. 9a,b) has a surface area of 97 m^2 g^{-1} and a total pore volume of 0.36 cm^3 g^{-1} and its adsorption capacities for As, Pb and Congo red are 57, 80 and 270 mg g^{-1}, respectively.

4.2 Regeneration of adsorbent using nanocatalysts

While adsorption technology is a versatile, economical and reliable technology for water remediation, the spent adsorbents are a hazardous waste which needs to be handled carefully. As an alternative to disposal, the adsorbents can be regenerated for reuse to reduce its associated disposal cost and landfill space. For AC (activated carbon), high-temperature thermal regeneration (up to 800°C) is the commonly adopted regeneration technique but it is energy-intensive, requires

Fig. 9. (a, c) and (b, d) are the FESEM and TEM images of the resulting product respectively. Sample A (a,b) was synthesized with the addition of 5 mL of glycerol and the total glycerol/water volume is fixed at 40 mL. Sample B (c,d) was synthesized with 3 mL of glycerol in the reaction solution.[242]

off-site regeneration, and incurs appreciable carbon loss (5–15%). Regeneration of the spent AC using solvents has associated drawbacks including solvent consumption and the generation of secondary waste streams. Solar photocatalytic regeneration of the AC which had been decorated with N-TiO$_2$ nanoparticles had been explored by NTU researchers.[243] It does not entail carbon footprint and chemical consumption during the adsorbent regeneration, and photocatalysis can non-selectively degrade various recalcitrant organic pollutants and the by-products are usually more environmentally benign. The regeneration efficiency depends on the type of adsorbed pollutants (their affinity for the AC), the composition of the N-TiO$_2$/AC, light intensity, and temperature (Fig. 10).

The sulfate radical-regeneration of an adsorbent had also been demonstrated using CuFe$_2$O$_4$/AC as catalytic adsorbent and peroxymonosulfate as oxidant.[244] The optimum ratio of CuFe$_2$O$_4$:AC was 1:1.5 for the best AC regeneration performance. The main factors hindering a complete CuFe$_2$O$_4$/AC regeneration were reported as (i) adsorption irreversibility of the pollutant, and (ii) surface modification of AC by peroxymonosulfate. With a mild heat treatment (150 °C) after the catalytic regeneration process, 82% of the adsorption capacity of the composite can be restored.

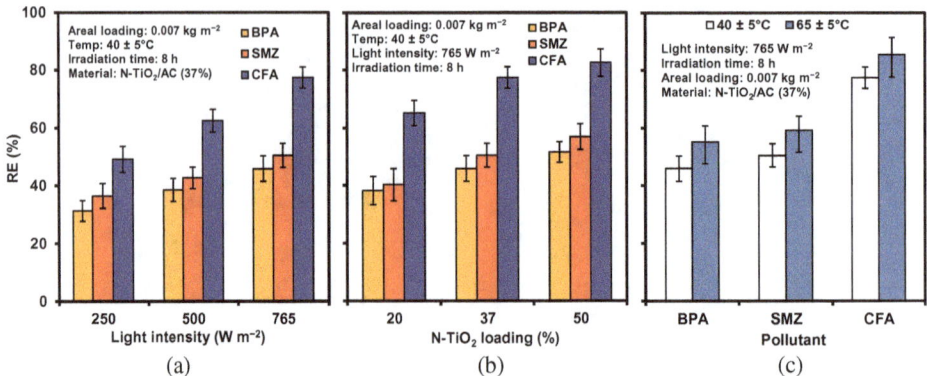

Fig. 10. The effects of (a) light intensity, (b) N-TiO$_2$ loading on the AC (wt%), and (c) temperature on solar photocatalytic regeneration efficiency of N-TiO$_2$/AC pre-saturated with bisphenol A (BPA), sulfamethazine (SMZ), and clofibric acid (CFA).

4.3 Synergistic coupling of adsorbent and photocatalyst

Knowing the respective advantages of photocatalyst and adsorbent, it is natural to combine the advantages of both for more effective water treatment. One of the early works was carried out by M.B. Ray et al.[245] who loaded different amounts of TiO$_2$ on inert adsorbent surfaces, including mesoporous (MCM-41), microporous (b-zeolite) and pillared structure (montmorillonite). All the supported catalysts exhibited good photodegradation efficiency of orange II, and their overall removal efficiency has always been better than that of bare TiO$_2$.

T.T. Lim's group[32] loaded TiO$_2$ nano particles on activated carbon and demonstrated the synergistic removal of Acid Red 88 via photocatalytic degradation and adsorption. His group has also recently demonstrated the synergistic adsorption-photocatalytic degradation of a hydrophobic pollutant under solar irradiation using a low temperature synthesized graphene/Bi$_2$Fe$_4$O$_9$ composite.[129]

Z. Chen's group fabricated *in situ* coupled photocatalyst/adsorbent materials via a one-pot solvothermal approach.[12,27,139] The prepared dual-phase particles possessed a high specific surface area (180 m^2 g^{-1}). A single particle has an average size around 20 nm, and consists of both anatase and titanate nanocrystallites that share an intimate interface. The synergized functioning of surface adsorption (titanate) and photo-degradation (sensitized anatase) has made the dual-phase particles much faster in degrading methylene blue dye under visible light illumination.[12,139] In another study, they prepared Ag-AgX (X = Cl, Br, I) nanoparticles anchored on a titanate nanotube (TNT) surface.[171] The Ag-AgX-TNT hybrid possesses a high surface area and could synergistically adsorb and degrade methylene blue under visible light.

5 Concluding Remarks and Future Perspectives

Current technologies for water remediation are generally not cost-effective nor energy-efficient in the production of clean water. The recent advance in nanotechnology and the emergence of a wide range of nanomaterials with diverse functionalities provide tremendous prospect for innovation and advancing technology for sustainable water remediation in the future. Nanomaterials having unique morphological, optical, electrical, and magnetic properties potentially provide breakthroughs in water treatment processes. This chapter summarizes a plethora of nanomaterials developed by Singapore's researchers for various applications in water remediation, including adsorption, oxidation, filtration, and disinfection. Despite the great advances made in materials synthesis and laboratory demonstration, most of the summarized works remain at the proof-of-concept or prototype stage. Most of the studies were conducted using synthetic water samples with conditions not representative of real water types. Future research should aim to develop the nanomaterials via scalable, economical and green synthesis techniques. Prototype devices or systems incorporating the nanomaterials for water remediation should be demonstrated on-site for treating real water samples.

References

1. Chen, D.W., and A.K. Ray, Photocatalytic kinetics of phenol and its derivatives over UV irradiated TiO2. *Appl. Catal. B-Environ*, 1999. **23**(2–3): pp. 143–157.
2. Qiao, S., D.D. Sun, J.H. Tay, and C. Easton, Photocatalytic oxidation technology for humic acid removal using a nano-structured TiO2/Fe2O3 catalyst. *Water Sci. Technol*, 2002. **47**(1): pp. 211–217.
3. Chiang, K., R. Amal, and T. Tran, Photocatalytic oxidation of cyanide: kinetic and mechanistic studies. *J. Mol. Catal. A-Chem*, 2003. **193**(1–2): pp.285–297.
4. Zhou, S.H., and A.K. Ray, Kinetic studies for photocatalytic degradation of eosin B on a thin film of titanium dioxide. *Ind. Eng. Chem. Res*, 2003. **42**(24): pp. 6020–6033.
5. Lin, G.F., J.W. Zheng, and R. Xu, Template-free synthesis of uniform CdS hollow nanospheres and their photocatalytic activities. *Journal of Physical Chemistry C*, 2008. **112**(19): pp. 7363–7370.
6. Yang, L., L.E. Yu, and M.B. Ray, Degradation of paracetamol in aqueous solutions by TiO2 photocatalysis. *Water Res*, 2008. **42**(13): pp. 3480–3488.
7. Zhang, X.W., A.J. Du, P.F. Lee, D.D. Sun, and J.O. Leckie, TiO2 nanowire membrane for concurrent filtration and photocatalytic oxidation of humic acid in water. *Journal of Membrane Science*, 2008. **313**(1–2): pp. 44–51.
8. Zhang, X.W., J.H. Pan, A.J. Du, P.F. Lee, D.D. Sun, and J.O. Lee, Aggregating TiO2 (B) nanowires to porous basketry-like microspheres and their photocatalytic properties. *Chem. Lett*, 2008. **37**(4): pp. 424–425.

9. Zhang, X.W., J.H. Pan, A.J. Du, W.J. Fu, D.D. Sun, and J.O. Leckie, Combination of one-dimensional TiO2 nanowire photocatalytic oxidation with microfiltration for water treatment. *Water Res*, 2009. **43**(5): pp. 1179–1186.
10. Li, G., L. Lv, H.T. Fan, J.Y. Ma, Y.Q. Li, Y. Wan, and X.S. Zhao, Effect of the agglomeration of TiO2 nanoparticles on their photocatalytic performance in the aqueous phase. *Journal of Colloid and Interface Science*, 2010. **348**(2): pp. 342–347.
11. Y.X. Zhang, B. Gao, G.L. Puma, A.K. Ray, and H.C. Zeng, Self-Assembled Au/TiO2/CNTs Ternary Nanocomposites for Photocatalytic Applications. *Science of Advanced Materials*, 2010. **2**(4): pp. 503–513.
12. Cheng, Y.H., Y.Z. Huang, P.D. Kanhere, V.P. Subramaniam, D.G. Gong, S. Zhang, J. Highfield, M.K. Schreyer, and Z. Chen, Dual-Phase Titanate/Anatase with Nitrogen Doping for Enhanced Degradation of Organic Dye under Visible Light. *Chemistry-a European Journal*, 2011. **17**(9): pp. 2575–2578.
13. Han, H., and R.B. Bai, Effect of Thickness of Photocatalyst Film Immobilized on a Buoyant Substrate on the Degradation of Methyl Orange Dye in Aqueous Solutions under Different Light Irradiations. *Ind. Eng. Chem. Res*, 2011. **50**(21): pp. 11922–11929.
14. Gong, D.G., W.C.J. Ho, Y.X. Tang, Q. Tay, Y.K. Lai, J.G. Highfield, and Z. Chen, Silver decorated titanate/titania nanostructures for efficient solar driven photocatalysis. *Journal of Solid State Chemistry*, 2012. **189**: 117–122.
15. Han, Z.A., V.W.C. Chang, L. Zhang, M.S. Tse, O.K. Tan, and L.M. Hildemann, Preparation of TiO2-Coated Polyester Fiber Filter by Spray-Coating and Its Photocatalytic Degradation of Gaseous Formaldehyde. *Aerosol and Air Quality Research*, 2012. **12**(6): pp. 1327–1335.
16. Tang, Y.X., P.X. Wee, Y.K. Lai, X.P. Wang, D.G. Gong, P.D. Kanhere, T.T. Lim, Z.L. Dong, and Z. Chen, Hierarchical TiO2 Nanoflakes and Nanoparticles Hybrid Structure for Improved Photocatalytic Activity. *Journal of Physical Chemistry C* 2012. **116**(4): pp. 2772–2780.
17. Xiong, P., and J.Y. Hu, Degradation of acetaminophen by UVA/LED/TiO2 process. *Separation and Purification Technology* 2012. **91**: 89–95.
18. Zhang, T., Y.J. Wang, J.W. Ng, and D.D. Sun, A free-standing, hybrid TiO2/K-OMS-2 hierarchical nanofibrous membrane with high photocatalytic activity for concurrent membrane filtration applications. *Rsc Advances* 2012. **2**(9): pp. 3638–3641.
19. Bai, H.W., L. Liu, Z.Y. Liu, and D.D. Sun, Hierarchical 3D dendritic TiO2 nanospheres building with ultralong 1D nanoribbon/wires for high performance concurrent photocatalytic membrane water purification. *Water Res*, 2013. **47**(12): pp. 4126–4138.
20. Gao, P., Z.Y. Liu, M.H. Tai, D.D. Sun, and W. Ng, Multifunctional graphene oxide-TiO2 microsphere hierarchical membrane for clean water production. *Appl. Catal. B-Environ*, 2013. **138**: 17–25.
21. Gao, P., D.D. Sun, and W.J. Ng, Multifunctional nanostructured membrane for clean water reclamation from wastewater with various pH conditions. *Rsc Advances* 2013. **3**(35): pp. 15202–15210.

22. Lee, S.S., H.W. Bai, Z.Y. Liu, and D.D. Sun, Novel-structured electrospun TiO2/CuO composite nanofibers for high efficient photocatalytic cogeneration of clean water and energy from dye wastewater. *Water Res*, 2013. **47**(12): pp. 4059–4073.
23. Miao, J.W., and B. Liu, Anatase TiO2 microspheres with reactive {001} facets for improved photocatalytic activity. *Rsc Advances* 2013. **3**(4): pp. 1222–1226.
24. Murugan, R., V.J. Babu, M.M. Khin, A.S. Nair, and S. Ramakrishna, Synthesis and photocatalytic applications of flower shaped electrospun ZnO-TiO2 mesostructures. *Materials Letters* 2013. **97**: 47–51.
25. Ong, W.L., M. Gao, and G.W. Ho, Hybrid organic PVDF-inorganic M-rGO-TiO2 (M = Ag, Pt) nanocomposites for multifunctional volatile organic compound sensing and photocatalytic degradation-H-2 production. *Nanoscale* 2013. **5**(22): pp. 11283–11290.
26. Zhou, W.J., Z.Y. Yin, Y.P. Du, X. Huang, Z.Y. Zeng, Z.X. Fan, H. Liu, J.Y. Wang, and H. Zhang, Synthesis of Few-Layer MoS2 Nanosheet-Coated TiO2 Nanobelt Heterostructures for Enhanced Photocatalytic Activities. *Small* 2013. **9**(1): pp. 140–147.
27. Cheng, Y.H., D.G. Gong, Y.X. Tang, J.W.C. Ho, Y.Y. Tay, W.S. Lau, O. Wijaya, J. Lim, and Z. Chen, One-pot solvothermal synthesis of dual-phase titanate/titania Nanoparticles and their adsorption and photocatalytic Performances. *Journal of Solid State Chemistry* 2014. **214**: pp. 67–73.
28. Guo, Z.G., J.K. Cheng, Z.G. Hu, M. Zhang, Q. Xu, Z.X. Kang, and D. Zhao, Metal-organic frameworks (MOFs) as precursors towards TiOx/C composites for photodegradation of organic dye. *Rsc Advances* 2014. **4**(65): pp. 34221–34225.
29. Liang, K., B.K. Tay, O.V. Kupreeva, T.I. Orekhoyskaya, S.K. Lazarouk, and V.E. Borisenko, Fabrication of Double-Walled Titania Nanotubes and Their Photocatalytic Activity. *Acs Sustainable Chemistry & Engineering* 2014. **2**(4): pp. 991–995.
30. Lin, X.H., and S.F.Y. Li, Determination of organic pollutants in municipal reverse osmosis concentrate by electrospray ionization-quadrupole time-of-flight tandem mass spectrometry and photocalaytic degradation methods. *Desalination* 2014. **344**: pp. 206–211.
31. Zhang, J.Y., F.X. Xiao, G.C. Xiao, and B. Liu, Assembly of a CdS quantum dot-TiO2 nanobelt heterostructure for photocatalytic application: towards an efficient visible light photocatalyst via facile surface charge tuning. *New Journal of Chemistry* 2015. **39**(1): pp. 279–286.
32. Gao, B., P.S. Yap, T.M. Lim, and T.T. Lim, Adsorption-photocatalytic degradation of Acid Red 88 by supported TiO2: Effect of activated carbon support and aqueous anions. *Chemical Engineering Journal* 2011. **171**(3): pp. 1098–1107.
33. Goei, R., Z. Dong, and T.T. Lim, High-permeability pluronic-based TiO2 hybrid photocatalytic membrane with hierarchical porosity: Fabrication, characterizations and performances. *Chemical Engineering Journal* 2013. **228**: pp. 1030–1039.
34. Liu, L., H. Bai, J. Liu, and D.D. Sun, Multifunctional graphene oxide-TiO_2-Ag nanocomposites for high performance water disinfection and decontamination under solar irradiation. *Journal of Hazardous Materials* 2013. **261**: pp. 214–223.

35. Wang, X., and T.T. Lim, Highly efficient and stable Ag-AgBr/TiO2 composites for destruction of Escherichia coli under visible light irradiation. *Water Res*, 2013. **47**(12): pp. 4148–4158.
36. Wang, X.L., S.O. Pehkonen, and A.K. Ray, Removal of aqueous Cr(VI) by a combination of photocatalytic reduction and coprecipitation. *Ind. Eng. Chem. Res*, 2004. **43**(7): pp. 1665–1672.
37. Wang, X.L., S.O. Pehkonen, and A.K. Ray, Photocatalytic reduction of Hg(II) on two commercial TiO2 catalysts. *Electrochim. Acta* 2004. **49**(9–10): pp.1435–1444.
38. Dutta, P.K., S.O. Pehkonen, V.K. Sharma, and A.K. Ray, Photocatalytic oxidation of arsenic(III): Evidence of hydroxyl radicals. *Environ. Sci. Technol*, 2005. **39**(6): pp. 1827–1834.
39. Zhang, X.W., T. Zhang, J.W. Ng, J.H. Pan, and D.D. Sun, Transformation of Bromine Species in TiO2 Photocatalytic System. *Environ. Sci. Technol*, 2010. **44**(1): pp. 439–444.
40. Xiao, F.X., S.F. Hung, H.B. Tao, J.W. Miao, H.B. Yang, and B. Liu, Spatially branched hierarchical ZnO nanorod-TiO2 nanotube array heterostructures for versatile photocatalytic and photoelectrocatalytic applications: towards intimate integration of 1D-1D hybrid nanostructures. *Nanoscale* 2014. **6**(24): pp. 14950–14961.
41. Zhu, L.L., M.H. Hong, and G.W. Ho, Hierarchical Assembly of SnO2/ZnO Nanostructures for Enhanced Photocatalytic Performance. *Scientific Reports* 2015. **5**: p, 11.
42. Highfield, J.G., M.H. Chen, P.T. Nguyen, and Z. Chen, Mechanistic investigations of photo-driven processes over TiO2 by in-situ DRIFTS-MS: Part 1. Platinization and methanol reforming. *Energy & Environmental Science* 2009. **2**(9): pp. 991–1002.
43. Gong, D.G., V.P. Subramaniam, J.G. Highfield, Y.X. Tang, Y.K. Lai, and Z. Chen, In Situ Mechanistic Investigation at the Liquid/Solid Interface by Attenuated Total Reflectance FTIR: Ethanol Photo-Oxidation over Pristine and Platinized TiO2 (P25). *Acs Catalysis* 2011. **1**(8): pp. 864–871.
44. Li, W., C. Liu, Y.X. Zhou, Y. Bai, X. Feng, Z.H. Yang, L.H. Lu, X.H. Lu, and K.Y. Chan, Enhanced Photocatalytic Activity in Anatase/TiO2(B) Core-Shell Nanofiber. *Journal of Physical Chemistry C* 2008. **112**(51): pp. 20539–20545.
45. Yang, D.J., H.W. Liu, Z.F. Zheng, Y. Yuan, J.C. Zhao, E.R. Waclawik, X.B. Ke, and H.Y. Zhu, An Efficient Photocatalyst Structure: TiO2(B) Nanofibers with a Shell of Anatase Nanocrystals. *Journal of the American Chemical Society* 2009. **131**(49): pp. 17885–17893.
46. Fu, N., Y.Q. Wu, Z.L. Jin, and G.X. Lu, Structural-Dependent Photoactivities of TiO2 Nanoribbon for Visible-Light-Induced H-2 Evolution: The Roles of Nanocavities and Alternate Structures. *Langmuir* 2010. **26**(1): pp. 447–455.
47. Zheng, Z.F., H.W. Liu, J.P. Ye, J.C. Zhao, E.R. Waclawik, and H.Y. Zhu, Structure and contribution to photocatalytic activity of the interfaces in nanofibers with mixed anatase and TiO2(B) phases. *J. Mol. Catal. A-Chem*, 2010. **316**(1–2): pp.75–82.
48. Ohtani, B., J. Handa, S. Nishimoto, and T. Kagiya, Highly-Active Semiconductor Photocatalyst - Extra-Fine Crystallite of Brookite Tio2 for Redox Reaction in Aqueous

Propan-2-Ol and or Silver Sulfate-Solution. *Chemical Physics Letters* 1985. **120**(3): pp. 292–294.

49. Kandiel, T.A., A. Feldhoff, L. Robben, R. Dillert, and D.W. Bahnemann, Tailored Titanium Dioxide Nanomaterials: Anatase Nanoparticles and Brookite Nanorods as Highly Active Photocatalysts. *Chem. Mat*, 2010. **22**(6): pp. 2050–2060.

50. Tay, Q.L., X.F. Liu, Y.X. Tang, Z.L. Jiang, T.C. Sum, and Z. Chen, Enhanced Photocatalytic Hydrogen Production with Synergistic Two-Phase Anatase/Brookite TiO2 Nanostructures. *Journal of Physical Chemistry C* 2013. **117**(29): pp. 14973–14982.

51. Han, M.Y., L.M. Gan, W. Huang, C.H. Chew, and B.S. Zou, Enhancement of photocatalytic oxidation activity by surface-modified CdS nanoparticles of high photostability. *Chem. Lett*, 1997. (8): pp. 751–752.

52. Zhang, W., Z.Y. Zhong, Y.S. Wang, and R. Xu, Doped Solid Solution: (Zn0.95Cu0.05)(1-x)CdxS Nanocrystals with High Activity for H-2 Evolution from Aqueous Solutions under Visible Light. *Journal of Physical Chemistry C* 2008. **112**(45): pp. 17635–17642.

53. Zhang, W., and R. Xu, Surface engineered active photocatalysts without noble metals: CuS-ZnxCd1-xS nanospheres by one-step synthesis. *International Journal of Hydrogen Energy* 2009. **34**(20): pp. 8495–8503.

54. Zhang, W., Y.B. Wang, Z. Wang, Z.Y. Zhong, and R. Xu, Highly efficient and noble metal-free NiS/CdS photocatalysts for H-2 evolution from lactic acid sacrificial solution under visible light. *Chemical Communications* 2010. **46**(40): pp. 7631–7633.

55. M.Y. Han, W. Huang, C.H. Quek, L.M. Gan, C.H. Chew, G.Q. Xu, and S.C. Ng, Preparation and enhanced photocatalytic oxidation activity of surface-modified CdS nanoparticles with high photostability. *J. Mater. Res*, 1999. **14**(5): pp. 2092–2095.

56. Gao, P., J.C. Liu, S. Lee, T. Zhang, and D.D. Sun, High quality graphene oxide-CdS-Pt nanocomposites for efficient photocatalytic hydrogen evolution. *Journal of Materials Chemistry* 2012. **22**(5): pp. 2292–2298.

57. Gao, P., J.C. Liu, T. Zhang, D.D. Sun, and W. Ng, Hierarchical TiO2/CdS "spindle-like" composite with high photodegradation and antibacterial capability under visible light irradiation. *Journal of Hazardous Materials* 2012, **229**: pp. 209–216.

58. Luo, J.S., L. Ma, T.C. He, C.F. Ng, S.J. Wang, H.D. Sun, and H.J. Fan, TiO2/(CdS, CdSe, CdSeS) Nanorod Heterostructures and Photoelectrochemical Properties. *Journal of Physical Chemistry C* 2012. **116**(22): pp. 11956–11963.

59. Cao, S.W., Y.P. Yuan, J. Fang, M.M. Shahjamali, F.Y.C. Boey, J. Barber, S.C.J. Loo, and C. Xue, In-situ growth of CdS quantum dots on g-C3N4 nanosheets for highly efficient photocatalytic hydrogen generation under visible light irradiation. *International Journal of Hydrogen Energy* 2013. **38**(3): pp. 1258–1266.

60. Fang, J., L. Xu, Z.Y. Zhang, Y.P. Yuan, S.W. Cao, Z. Wang, L.S. Yin, Y.S. Liao, and C. Xue, Au@TiO2-CdS Ternary Nanostructures for Efficient Visible-Light-Driven Hydrogen Generation. *Acs Applied Materials & Interfaces* 2013. **5**(16): pp. 8088–8092.

61. Hu, Y., X.H. Gao, L. Yu, Y.R. Wang, J.Q. Ning, S.J. Xu, and X.W. Lou, Carbon-Coated CdS Petalous Nanostructures with Enhanced Photostability and Photocatalytic Activity. *Angew. Chem.-Int. Edit*, 2013. **52**(21): pp. 5636–5639.

62. Wang, Y.B., Y.S. Wang, and R. Xu, Photochemical Deposition of Pt on CdS for H-2 Evolution from Water: Markedly Enhanced Activity by Controlling Pt Reduction Environment. *Journal of Physical Chemistry C* 2013. **117**(2): pp. 783–790.
63. Xiao, F.X., J.W. Miao, and B. Liu, Layer-by-Layer Self-Assembly of CdS Quantum Dots/Graphene Nanosheets Hybrid Films for Photoelectrochemical and Photocatalytic Applications. *Journal of the American Chemical Society* 2014. **136**(4): pp. 1559–1569.
64. Xiao, F.X., J.W. Miao, and B. Liu, Self-assembly of aligned rutile@anatase TiO2 nanorod@CdS quantum dots ternary core-shell heterostructure: cascade electron transfer by interfacial design. *Materials Horizons* 2014. **1**(2): pp. 259–263.
65. Chen, J.Z., X.J. Wu, L.S. Yin, B. Li, X. Hong, Z.X. Fan, B. Chen, C. Xue, and H. Zhang, One-pot Synthesis of CdS Nanocrystals Hybridized with Single-Layer Transition-Metal Dichalcogenide Nanosheets for Efficient Photocatalytic Hydrogen Evolution. *Angew. Chem.-Int. Edit*, 2015. **54**(4): pp. 1210–1214.
66. Dong, W.B., F. Pan, L.L. Xu, M.R. Zheng, C.H. Sow, K. Wu, G.Q. Xu, and W. Chen, Facile synthesis of CdS@TiO2 core-shell nanorods with controllable shell thickness and enhanced photocatalytic activity under visible light irradiation. *Applied Surface Science* 2015. **349**: pp. 279–286.
67. Gao, P., J. Liu, D.D. Sun, and W. Ng, Graphene oxide-CdS composite with high photocatalytic degradation and disinfection activities under visible light irradiation. *Journal of Hazardous Materials* 2013. 250–251: pp. 412–420.
68. Yang, H.Y., S.F. Yu, S.P. Lau, X.W. Zhang, D.D. Sun, and G. Jun, Direct Growth of ZnO Nanocrystals onto the Surface of Porous TiO2 Nanotube Arrays for Highly Efficient and Recyclable Photocatalysts. *Small* 2009. **5**(20): pp. 2260–2264.
69. Yuan, J.Q., E.S.G. Choo, X.S. Tang, Y. Sheng, J. Ding, and J.M. Xue, Synthesis of ZnO-Pt nanoflowers and their photocatalytic applications. *Nanotechnology* 2010. **21**(18): p. 10.
70. Bai, H.W., Z.Y. Liu, and D.D. Sun, Hierarchical ZnO/Cu "corn-like" materials with high photodegradation and antibacterial capability under visible light. *Physical Chemistry Chemical Physics*, 2011. **13**(13): pp. 6205–6210.
71. Liu, Z.Y., H.W. Bai, S.P. Xu, and D.D. Sun, Hierarchical CuO/ZnO "corn-like" architecture for photocatalytic hydrogen generation. *International Journal of Hydrogen Energy*, 2011. **36**(21): pp. 13473–13480.
72. Bai, H.W., Z.Y. Liu, and D.D. Sun, Hierarchical ZnO nanostructured membrane for multifunctional environmental applications. *Colloids and Surfaces a-Physicochemical and Engineering Aspects*, 2012. **410**: pp. 11–17.
73. Bai, H.W., Z.Y. Liu, and D.D. Sun, Solar-Light-Driven Photodegradation and Antibacterial Activity of Hierarchical TiO2/ZnO/CuO Material. *Chempluschem*, 2012. **77**(10): pp. 941–948.
74. Liu, Z.Y., H.W. Bai, and D.D. Sun, Hierarchical CuO/ZnOMembranes for Environmental Applications under the Irradiation of Visible Light. *International Journal of Photoenergy*, 2012. p. 11.
75. Pan, J.H., X.W. Zhang, A.J. Du, H.W. Bai, J.W. Ng, and D.R. Sun, A hierarchically assembled mesoporous ZnO hemisphere array and hollow microspheres for photocatalytic

membrane water filtration. *Physical Chemistry Chemical Physics*, 2012. **14**(20): pp. 7481–7489.
76. Wang, Y.B., Y.S. Wang, R.R. Jiang, and R. Xu, Cobalt Phosphate-ZnO Composite Photocatalysts for Oxygen Evolution from Photocatalytic Water Oxidation. *Ind. Eng. Chem. Res*, 2012. **51**(30): pp. 9945–9951.
77. Wang, Y.J., J.C. Liu, L. Liu, and D.D. Sun, Enhancing Stability and Photocatalytic Activity of ZnO Nanoparticles by Surface Modification of Graphene Oxide. *Journal of Nanoscience and Nanotechnology*, 2012. **12**(5): pp. 3896–3902.
78. Wei, Y.F., L. Ke, E.S.P. Leong, H. Liu, L.L. Liew, J.H. Teng, H.J. Du, and X.W. Sun, Enhanced photoelectrochemical performance of bridged ZnO nanorod arrays grown on V-grooved structure. *Nanotechnology*, 2012. **23**(36): p. 6.
79. Gao, P., K. Ng, and D.D. Sun, Sulfonated graphene oxide-ZnO-Ag photocatalyst for fast photodegradation and disinfection under visible light. *Journal of Hazardous Materials*, 2013. **262**: pp. 826–835.
80. Liu, Z.Y., H.W. Bai, and D.D. Sun, Hierarchical ZnO Nanoflake Structured Multifunctional Membrane for Water Purification. *Separation Science and Technology*, 2013. **48**(3): pp. 473–479.
81. Liu, Z.Y., H.W. Bai, and D.D. Sun, A general method for the fabrication of hierarchically-nanostructured membranes with multifunctional environmental applications. *Separation and Purification Technology*, 2013. **107**: pp. 324–330.
82. Ong, W.L., S. Natarajan, B. Kloostra, and G.W. Ho, Metal nanoparticle-loaded hierarchically assembled ZnO nanoflakes for enhanced photocatalytic performance. *Nanoscale*, 2013. **5**(12): pp. 5568–5575.
83. Ong, W.L., K.W. Yew, C.F. Tan, T.K.T. Adrian, M.H. Hong, and G.W. Ho, Highly flexible solution processable heterostructured zinc oxide nanowires mesh for environmental clean-up applications. *Rsc Advances*, 2014. **4**(52): pp. 27481–27487.
84. Zhang, X.L., Y. Li, J.L. Zhao, S.G. Wang, Y.D. Li, H.T. Dai, and X.W. Sun, Advanced three-component ZnO/Ag/CdS nanocomposite photoanode for photocatalytic water splitting. *Journal of Power Sources*, 2014. **269**: pp. 466–472.
85. Bai, H., Z. Liu, and D.D. Sun, A hierarchically structured and multifunctional membrane for water treatment. *Applied Catalysis B: Environmental*, 2012. **111–112**(0): pp. 571–577.
86. Hong, J.D., X.Y. Xia, Y.S. Wang, and R. Xu, Mesoporous carbon nitride with in situ sulfur doping for enhanced photocatalytic hydrogen evolution from water under visible light. *Journal of Materials Chemistry*, 2012. **22**(30): pp. 15006–15012.
87. Cao, S.W., X.F. Liu, Y.P. Yuan, Z.Y. Zhang, J. Fang, S.C.J. Loo, J. Barber, T.C. Sum, and C. Xue, Artificial photosynthetic hydrogen evolution over g-C3N4 nanosheets coupled with cobaloxime. *Physical Chemistry Chemical Physics*, 2013. **15**(42): pp. 18363–18366.
88. Yuan, Y.P., S.W. Cao, Y.S. Liao, L.S. Yin, and C. Xue, Red phosphor/g-C3N4 heterojunction with enhanced photocatalytic activities for solar fuels production. *Appl. Catal. B-Environ,*, 2013. **140**: pp. 164–168.
89. Hong, J.D., S.M. Yin, Y.X. Pan, J.Y. Han, T.H. Zhou, and R. Xu, Porous carbon nitride nanosheets for enhanced photocatalytic activities. *Nanoscale*, 2014. **6**(24): pp. 14984–14990.

90. Hong, J.D., W. Zhang, Y.B. Wang, T.H. Zhou, and R. Xu, Photocatalytic Reduction of Carbon Dioxide over Self-Assembled Carbon Nitride and Layered Double Hydroxide: The Role of Carbon Dioxide Enrichment. *Chemcatchem*, 2014. **6**(8): pp. 2315–2321.
91. Yin, L.S., Y.P. Yuan, S.W. Cao, Z.Y. Zhang, and C. Xue, Enhanced visible-light-driven photocatalytic hydrogen generation over g-C3N4 through loading the noble metal-free NiS2 cocatalyst. *Rsc Advances*, 2014. **4**(12): pp. 6127–6132.
92. Gu, Q., Z.W. Gao, H.G. Zhao, Z.Z. Lou, Y.S. Liao, and C. Xue, Temperature-controlled morphology evolution of graphitic carbon nitride nanostructures and their photocatalytic activities under visible light. *Rsc Advances*, 2015. **5**(61): pp. 49317–49325.
93. Rahimnejad, S., G.Q. Xu, and W. Chen, Visible WO3 photocatalyst with tunable band gap for energy conversion. *Abstracts of Papers of the American Chemical Society*, 2014. **247**: p. 1.
94. Lou, Z.Z., Q. Gu, L. Xu, Y.S. Liao, and C. Xue, Surfactant-Free Synthesis of Plasmonic Tungsten Oxide Nanowires with Visible-Light-Enhanced Hydrogen Generation from Ammonia Borane. *Chemistry-an Asian Journal*, 2015. **10**(6): pp. 1291–1294.
95. Qi, H., J. Wolfe, D.P. Wang, H.J. Fan, D. Fichou, and Z. Chen, Triple-layered nanostructured WO3 photoanodes with enhanced photocurrent generation and superior stability for photoelectrochemical solar energy conversion. *Nanoscale*, 2014. **6**(22): pp. 13457–13462.
96. Booshehri, A.Y., S.C.K. Goh, J.D. Hong, R.R. Jiang, and R. Xu, Effect of depositing silver nanoparticles on BiVO4 in enhancing visible light photocatalytic inactivation of bacteria in water. *Journal of Materials Chemistry A*, 2014. **2**(17): pp. 6209–6217.
97. Cao, S.W., Z. Yin, J. Barber, F.Y.C. Boey, S.C.J. Loo, and C. Xue, Preparation of Au-BiVO4 Heterogeneous Nanostructures as Highly Efficient Visible-Light Photocatalysts. *Acs Applied Materials & Interfaces*, 2012. **4**(1): pp. 418–423.
98. Gao, X.H., Bin H.B. Wu, L.X. Zheng, Y.J. Zhong, Y. Hu, and X.W. Lou, Formation of Mesoporous Heterostructured BiVO4/Bi2S3 Hollow Discoids with Enhanced Photoactivity. *Angew. Chem.-Int. Edit,*, 2014. **53**(23): pp. 5917–5921.
99. Han, M.D., X.F. Chen, T. Sun, O.K. Tan, and M.S. Tse, Synthesis of mono-dispersed m-BiVO4 octahedral nano-crystals with enhanced visible light photocatalytic properties. *Crystengcomm*, 2011. **13**(22): pp. 6674–6679.
100. Han, M.D., T. Sun, P.Y. Tan, X.F. Chen, O.K. Tan, and M.S. Tse, m-BiVO4@gamma-Bi2O3 core-shell p-n heterogeneous nanostructure for enhanced visible-light photocatalytic performance. *Rsc Advances*, 2013. **3**(47): pp. 24964–24970.
101. Kanhere, P., J. Nisar, Y.X. Tang, B. Pathak, R. Ahuja, J.W. Zheng, and Z. Chen, Electronic Structure, Optical Properties, and Photocatalytic Activities of LaFeO3-NaTaO3 Solid Solution. *Journal of Physical Chemistry C*, 2012. **116**(43): pp. 22767–22773.
102. Kanhere, P., P. Shenai, S. Chakraborty, R. Ahuja, J.W. Zheng, and Z. Chen, Mono- and co-doped NaTaO3 for visible light photocatalysis. *Physical Chemistry Chemical Physics*, 2014. **16**(30): pp. 16085–16094.
103. Kanhere, P., Y.X. Tang, J.W. Zheng, and Z. Chen, Synthesis, photophysical properties, and photocatalytic applications of Bi doped NaTaO3 and Bi doped Na2Ta2O6 nanoparticles. *J. Phys. Chem. Solids*, 2013. **74**(12): pp. 1708–1713.

104. Kanhere, P., J.W. Zheng, and Z. Chen, Visible light driven photocatalytic hydrogen evolution and photophysical properties of Bi3+ doped NaTaO3. *International Journal of Hydrogen Energy*, 2012. **37**(6): pp. 4889–4896.
105. Kanhere, P.D., J.W. Zheng, and Z. Chen, Site Specific Optical and Photocatalytic Properties of Bi-Doped NaTaO3. *Journal of Physical Chemistry C*, 2011. **115**(23): pp. 11846–11853.
106. Wang, B.C., P.D. Kanhere, Z. Chen, J. Nisar, B. Pathak, and R. Ahuja, Anion-Doped NaTaO3 for Visible Light Photocatalysis. *Journal of Physical Chemistry C*, 2013. **117**(44): pp. 22518–22524.
107. Zhou, W.W., B. Yan, C.W. Cheng, C.X. Cong, H.L. Hu, H.J. Fan, and T. Yu, Facile synthesis and shape evolution of highly symmetric 26-facet polyhedral microcrystals of Cu2O. *Crystengcomm*, 2009. **11**(11): pp. 2291–2296.
108. Liu, Y., P.D. Kanhere, C.L. Wong, Y.F. Tian, Y.H. Feng, F. Boey, T. Wu, H.Y. Chen, T.J. White, Z. Chen, and Q.C. Zhang, Hydrazine-hydrothermal method to synthesize three-dimensional chalcogenide framework for photocatalytic hydrogen generation. *Journal of Solid State Chemistry*, 2010. **183**(11): pp. 2644–2649.
109. Pang, M.L., J.Y. Hu, and H.C. Zeng, Synthesis, Morphological Control, and Antibacterial Properties of Hollow/Solid Ag2S/Ag Heterodimers. *Journal of the American Chemical Society*, 2010. **132**(31): pp. 10771–10785.
110. Tan, K.Y.D., G.F. Teng, and W.Y. Fan, Photocatalytic Transformation of Organic and Water-Soluble Thiols into Disulfides and Hydrogen under Aerobic Conditions Using Mn(CO)(5)Br. *Organometallics*, 2011. **30**(15): pp. 4136–4143.
111. Zhang, T., X.W. Zhang, J.W. Ng, H.Y. Yang, J.C. Liu, and D.D. Sun, Fabrication of magnetic cryptomelane-type manganese oxide nanowires for water treatment. *Chemical Communications*, 2011. **47**(6): pp. 1890–1892.
112. Cao, S.W., J. Fang, M.M. Shahjamali, Z. Wang, Z. Yin, Y.H. Yang, F.Y.C. Boey, J. Barber, S.C.J. Loo, and C. Xue, In situ growth of Au nanoparticles on Fe2O3 nanocrystals for catalytic applications. *Crystengcomm*, 2012. **14**(21): pp. 7229–7235.
113. Liu, L., J.C. Liu, and D.D. Sun, Graphene oxide enwrapped Ag3PO4 composite: towards a highly efficient and stable visible-light-induced photocatalyst for water purification. *Catalysis Science & Technology*, 2012. **2**(12): pp. 2525–2532.
114. Pan, Y.L., S.Z. Deng, L. Polavarapu, N.Y. Gao, P.Y. Yuan, C.H. Sow, and Q.H. Xu, Plasmon-Enhanced Photocatalytic Properties of Cu2O Nanowire-Au Nanoparticle Assemblies. *Langmuir*, 2012. **28**(33): pp. 12304–12310.
115. Tran, P.D., S.K. Batabyal, S.S. Pramana, J. Barber, L.H. Wong, and S.C.J. Loo, A cuprous oxide-reduced graphene oxide (Cu2O-rGO) composite photocatalyst for hydrogen generation: employing rGO as an electron acceptor to enhance the photocatalytic activity and stability of Cu2O. *Nanoscale*, 2012. **4**(13): pp. 3875–3878.
116. Yang, W.L., L. Zhang, Y. Hu, Y.J. Zhong, H.B. Wu, and X.W. Lou, Microwave-Assisted Synthesis of Porous Ag2S-Ag Hybrid Nanotubes with High Visible-Light Photocatalytic Activity. *Angew. Chem.-Int. Edit*, 2012. **51**(46): pp. 11501–11504.

117. Chong, Y.Y., and W.Y. Fan, Facile Synthesis of Single Crystalline Rhenium (VI) Trioxide Nanocubes with High Catalytic Efficiency for Photodegradation of Methyl Orange. *Journal of Colloid and Interface Science*, 2013. **397**: pp. 18–23.
118. Peng, S.J., L.L. Li, Y.Z. Wu, L. Jia, L.L. Tian, M. Srinivasan, S. Ramakrishna, Q.Y. Yan, and S.G. Mhaisalkar, Size- and shape-controlled synthesis of ZnIn2S4 nanocrystals with high photocatalytic performance. *Crystengcomm*, 2013. **15**(10): pp. 1922–1930.
119. Peng, S.J., L.L. Li, P.N. Zhu, Y.Z. Wu, M. Srinivasan, S.G. Mhaisalkar, S. Ramakrishna, and Q.Y. Yan, Controlled Synthesis of BiOCl Hierarchical Self-Assemblies with Highly Efficient Photocatalytic Properties. *Chemistry-an Asian Journal*, 2013. **8**(1): pp. 258–268.
120. Regulacio, M.D., C. Ye, S.H. Lim, Y.G. Zheng, Q.H. Xu, and M.Y. Han, Facile noninjection synthesis and photocatalytic properties of wurtzite-phase CuGaS2 nanocrystals with elongated morphologies. *Crystengcomm*, 2013. **15**(26): pp. 5214–5217.
121. L. Yao, L.Z. Zhang, R. Wang, C.H. Loh, and Z.L. Dong, Fabrication of catalytic membrane contactors based on polyoxometalates and polyvinylidene fluoride intended for degrading phenol in wastewater under mild conditions. *Separation and Purification Technology*, 2013. **118**: pp. 162–169.
122. Babu, V.J., R.S.R. Bhavatharini, and S. Ramakrishna, Electrospun BiOI nano/microtectonic plate-like structure synthesis and UV-light assisted photodegradation of ARS dye. *Rsc Advances*, 2014. **4**(37): pp. 19251–19256.
123. Gao, J.K., J.W. Miao, P.Z. Li, W.Y. Teng, L. Yang, Y.L. Zhao, B. Liu, and Q.C. Zhang, A p-type Ti(IV)-based metal-organic framework with visible-light photo-response. *Chemical Communications*, 2014. **50**(29): pp. 3786–3788.
124. Gao, P., A.R. Li, M.H. Tai, Z.Y. Liu, and D.D. Sun, A Hierarchical Nanostructured Carbon Nanofiber-In2S3 Photocatalyst with High Photodegradation and Disinfection Abilities Under Visible Light. *Chemistry-an Asian Journal*, 2014. **9**(6): pp. 1663–1670.
125. Kanhere, P., and Z. Chen, A Review on Visible Light Active Perovskite-Based Photocatalysts. *Molecules*, 2014. **19**(12): pp. 19995–20022.
126. Muduli, S.K., S.L. Wang, S. Chen, C.F. Ng, C.H.A. Huan, T.C. Sum, and H.S. Soo, Mesoporous cerium oxide nanospheres for the visible-light driven photocatalytic degradation of dyes. *Beilstein Journal of Nanotechnology*, 2014. **5**: pp. 517–523.
127. Nie, L.N., W.W. Xiong, P.Z. Li, J.Y. Han, G.D. Zhang, S.M. Yin, Y.L. Zhao, R. Xu, and Q.C. Zhang, Surfactant-thermal method to prepare two novel two-dimensional Mn-Sb-S compounds for photocatalytic applications. *Journal of Solid State Chemistry*, 2014. **220**: pp. 118–123.
128. Sha, Z., J.L. Sun, H.S.O. Chan, S. Jaenicke, and J.S. Wu, Bismuth tungstate incorporated zirconium metalorganic framework composite with enhanced visible-light photocatalytic performance. *Rsc Advances*, 2014. **4**(110): pp. 64977–64984.
129. Hu, Z.T., J. Liu, X. Yan, W.D. Oh, and T.T. Lim, Low-temperature synthesis of graphene/Bi2Fe4O9 composite for synergistic adsorption-photocatalytic degradation of hydrophobic pollutant under solar irradiation. *Chemical Engineering Journal*, 2015. **262**: pp. 1022–1032.

130. Sha, Z., and J.S. Wu, Enhanced visible-light photocatalytic performance of BiOBr/UiO-66(Zr) composite for dye degradation with the assistance of UiO-66. *Rsc Advances*, 2015. **5**(49): pp. 39592–39600.
131. Chen, J.S., C.P. Chen, J. Liu, R. Xu, S.Z. Qiao, and X.W. Lou, Ellipsoidal hollow nanostructures assembled from anatase TiO2 nanosheets as a magnetically separable photocatalyst. *Chemical Communications*, 2011. **47**(9): pp. 2631–2633.
132. Liu, J.C., L. Liu, H.W. Bai, Y.J. Wang, and D.D. Sun, Gram-scale production of graphene oxide-TiO2 nanorod composites: Towards high-activity photocatalytic materials. *Appl. Catal. B-Environ*, 2011. **106**(1–2): pp.76–82.
133. Pan, J.H., G. Han, R. Zhou, and X.S. Zhao, Hierarchical N-doped TiO2 hollow microspheres consisting of nanothorns with exposed anatase {101} facets. *Chemical Communications*, 2011. **47**(24): pp. 6942–6944.
134. Bai, H.W., Z.Y. Liu, and D.D. Sun, Facile preparation of monodisperse, carbon doped single crystal rutile TiO2 nanorod spheres with a large percentage of reactive (110) facet exposure for highly efficient H-2 generation. *Journal of Materials Chemistry*, 2012. **22**(36): pp. 18801–18807.
135. Jiang, Z.L., Y.X. Tang, Q.L. Tay, Y.Y. Zhang, O.I. Malyi, D.P. Wang, J.Y. Deng, Y.K. Lai, H.F. Zhou, X.D. Chen, Z.L. Dong, and Z. Chen, Understanding the Role of Nanostructures for Efficient Hydrogen Generation on Immobilized Photocatalysts. *Advanced Energy Materials*, 2013. **3**(10): pp. 1368–1380.
136. Wang, D.P., P. Kanhere, M.J. Li, Q.L. Tay, Y.X. Tang, Y.Z. Huang, T.C. Sum, N. Mathews, T. Sritharan, and Z. Chen, Improving Photocatalytic H-2 Evolution of TiO2 via Formation of {001}-{010} Quasi-Heterojunctions. *Journal of Physical Chemistry C*, 2013. **117**(44): pp. 22894–22902.
137. Pan, F., K. Wu, H.X. Li, G.Q. Xu, and W. Chen, Synthesis of {100} Facet Dominant Anatase TiO2 Nanobelts and the Origin of Facet-Dependent Photoreactivity. *Chemistry-a European Journal*, 2014. **20**(46): pp. 15095–15101.
138. Ng, J., S.P. Xu, X.W. Zhang, H.Y. Yang, and D.D. Sun, Hybridized Nanowires and Cubes: A Novel Architecture of a Heterojunctioned TiO2/SrTiO3 Thin Film for Efficient Water Splitting. *Advanced Functional Materials*, 2010. **20**(24): pp. 4287–4294.
139. Cheng, Y.H., V.P. Subramaniam, D.G. Gong, Y.X. Tang, J. Highfield, S.O. Pehkonen, P. Pichat, M.K. Schreyer, and Z. Chen, Nitrogen-sensitized dual phase titanate/titania for visible-light driven phenol degradation. *Journal of Solid State Chemistry*, 2012. **196**: pp. 518–527.
140. Zhang, T., X.L. Yan, and D.D. Sun, Hierarchically multifunctional K-OMS-2/TiO2/Fe3O4 heterojunctions for the photocatalytic oxidation of humic acid under solar light irradiation. *Journal of Hazardous Materials*, 2012. **243**: pp. 302–310.
141. Lin, J.D., P. Hu, Y. Zhang, M.T. Fan, Z.M. He, C.K. Ngaw, J.S.C. Loo, D.W. Liao, and T.T.Y. Tan, Understanding the photoelectrochemical properties of a reduced graphene oxide-WO3 heterojunction photoanode for efficient solar-light-driven overall water splitting. *Rsc Advances*, 2013. **3**(24): pp. 9330–9336.

142. M.M. Gao, C.K.N. Peh, Y.L. Pan, Q.H. Xu, and G.W. Ho, Fine structural tuning of whereabout and clustering of metal-metal oxide heterostructure for optimal photocatalytic enhancement and stability. *Nanoscale*, 2014. **6**(21): pp. 12655–12664.
143. Bai, H.W., X.L. Zan, J. Juay, and D.D. Sun, Hierarchical heteroarchitectures functionalized membrane for high efficient water purification. *Journal of Membrane Science* 2015, **475**: pp. 245–251.
144. Wang, X.P., and T.T. Lim, Effect of hexamethylenetetramine on the visible-light photocatalytic activity of C–N codoped TiO_2 for bisphenol A degradation: evaluation of photocatalytic mechanism and solution toxicity *Applied Catalysis A: General*, 2011. **399**(1–2): pp.233–241.
145. Wang, X.P., and T.T. Lim, Solvothermal synthesis of C-N codoped TiO_2 and photocatalytic evaluation for bisphenol A degradation using a visible-light irradiated LED photoreactor. *Applied Catalysis B: Environmental*, 2010. **100**(1–2): pp.355–364.
146. Wang, P., T. Zhou, R. Wang, and T.T. Lim, Carbon-sensitized and nitrogen-doped TiO2 for photocatalytic degradation of sulfanilamide under visible-light irradiation. *Water Res,*, 2011. **45**(16): pp. 5015–5026.
147. Wang, P., A.G. Fane, and T.T. Lim, Evaluation of a submerged membrane vis-LED photoreactor (sMPR) for carbamazepine degradation and TiO2 separation. *Chemical Engineering Journal*, 2013. **215–216**: pp. 240–251.
148. Hou, D., R. Goei, X. Wang, P. Wang, and T.T. Lim, Preparation of carbon-sensitized and Fe-Er codoped TiO 2 with response surface methodology for bisphenol A photocatalytic degradation under visible-light irradiation. *Applied Catalysis B: Environmental*, 2012. **126**: pp. 121–133.
149. Yap, P.S., T.T. Lim, and M. Srinivasan, Nitrogen-doped TiO2/AC bi-functional composite prepared by two-stage calcination for enhanced synergistic removal of hydrophobic pollutant using solar irradiation. *Catal. Today*, 2011. **161**(1): pp. 46–52.
150. Yap, P.S., and T.T. Lim, Effect of aqueous matrix species on synergistic removal of bisphenol-A under solar irradiation using nitrogen-doped TiO2/AC composite. *Applied Catalysis B: Environmental*, 2011. **101**(3–4): pp.709–717.
151. Wang, P., and T.T. Lim, Membrane vis-LED photoreactor for simultaneous penicillin G degradation and TiO 2 separation. *Water Res*, 2012. **46**(6): pp. 1825–1837.
152. Wang, P., P.S. Yap, and T.T. Lim, C-N-S tridoped TiO2 for photocatalytic degradation of tetracycline under visible-light irradiation. *Applied Catalysis A: General*, 2011. **399**(1–2): pp.252–261.
153. Wang, X.P., Y.X. Tang, M.Y. Leiw, and T.T. Lim, Solvothermal synthesis of Fe-C codoped TiO_2 nanoparticles for visible-light photocatalytic removal of emerging organic contaminants in water. *Appl. Catal. A-Gen*, 2011. **409**: pp. 257–266.
154. Wang, P., B. Huang, Y. Dai, and M.H. Whangbo, Plasmonic photocatalysts: Harvesting visible light with noble metal nanoparticles. *Physical Chemistry Chemical Physics*, 2012. **14**(28): pp. 9813–9825.
155. Serpone, N., and A.V. Emeline, Semiconductor photocatalysis - Past, present, and future outlook. *Journal of Physical Chemistry Letters*, 2012. **3**(5): pp. 673–677.

156. Rycenga, M., C.M. Cobley, J. Zeng, W. Li, C.H. Moran, Q. Zhang, D. Qin, and Y. Xia, Controlling the synthesis and assembly of silver nanostructures for plasmonic applications. *Chemical Reviews*, 2011. **111**(6): pp. 3669–3712.
157. Zhou, X., G. Liu, J. Yu, and W. Fan, Surface plasmon resonance-mediated photocatalysis by noble metal-based composites under visible light. *Journal of Materials Chemistry*, 2012. **22**(40): pp. 21337–21354.
158. Zielińska-Jurek, A., E. Kowalska, J.W. Sobczak, W. Lisowski, B. Ohtani, and A. Zaleska, Preparation and characterization of monometallic (Au) and bimetallic (Ag/Au) modified-titania photocatalysts activated by visible light. *Applied Catalysis B: Environmental*, 2011. **101**(3–4): pp.504–514.
159. Konishi, Y., I. Tanabe, and T. Tatsuma, Stable spectral dip formation and multicolour changes of plasmonic gold nanoparticles on TiO_2. *Chemical Communications*, 2013. **49**(6): pp. 606–608.
160. Barka-Bouaifel, F., K. Makaoui, P.Y. Jouan, X. Castel, N. Bezzi, R. Boukherroub, and S. Szunerits, Preparation and photocatalytic properties of quartz/gold nanostructures/TiO_2 lamellar structures. *RSC Advances*, 2012. **2**(32): pp. 12482–12488.
161. Nishijima, Y., K. Ueno, Y. Yokota, K. Murakoshi, and H. Misawa, Plasmon-assisted photocurrent generation from visible to near-infrared wavelength using a Au-nanorods/TiO_2 electrode. *Journal of Physical Chemistry Letters*, 2010. **1**(13): pp. 2031–2036.
162. Li, X., S. Fang, L. Ge, C. Han, P. Qiu, and W. Liu, Synthesis of flower-like Ag/AgCl-Bi_2MoO_6 plasmonic photocatalysts with enhanced visible-light photocatalytic performance. *Applied Catalysis B: Environmental*, 2015. **176–177**: pp. 62–69.
163. Zhang, Z., A. Li, S.W. Cao, M. Bosman, S. Li, and C. Xue, Direct evidence of plasmon enhancement on photocatalytic hydrogen generation over Au/Pt-decorated TiO2 nanofibers. *Nanoscale*, 2014. **6**(10): pp. 5217–5222.
164. Tanaka, A., K. Hashimoto, and H. Kominami, Visible-light-induced hydrogen and oxygen formation over Pt/Au/WO_3 photocatalyst utilizing two types of photoabsorption due to surface plasmon resonance and band-gap excitation. *Journal of the American Chemical Society*, 2014. **136**(2): pp. 586–589.
165. Peng, Z.P., H.L. Hu, M.I.B. Utama, L.M. Wong, K. Ghosh, R.J. Chen, S.J. Wang, Z.X. Shen, and Q.H. Xiong, Heteroepitaxial Decoration of Ag Nanoparticles on Si Nanowires: A Case Study on Raman Scattering and Mapping. *Nano Letters*, 2010. **10**(10): pp. 3940–3947.
166. Kumar, M.K., S. Krishnamoorthy, L.K. Tan, S.Y. Chiam, S. Tripathy, and H. Gao, Field Effects in Plasmonic Photocatalyst by Precise SiO2 Thickness Control Using Atomic Layer Deposition. *Acs Catalysis*, 2011. **1**(4): pp. 300–308.
167. Tang, Y.X., V.P. Subramaniam, T.H. Lau, Y.K. Lai, D.G. Gong, P.D. Kanhere, Y.H. Cheng, Z. Chen, and Z.L. Dong, In situ formation of large-scale Ag/AgCl nanoparticles on layered titanate honeycomb by gas phase reaction for visible light degradation of phenol solution. *Appl. Catal. B-Environ*, 2011. **106**(3–4): pp. 577–585.
168. Fang, J., S.W. Cao, Z. Wang, M.M. Shahjamali, S.C.J. Loo, J. Barber, and C. Xue, Mesoporous plasmonic Au-TiO2 nanocomposites for efficient visible-light-driven

photocatalytic water reduction. *International Journal of Hydrogen Energy*, 2012. **37**(23): pp. 17853–17861.
169. Seh, Z.W., S.H. Liu, M. Low, S.Y. Zhang, Z.L. Liu, A. Mlayah, and M.Y. Han, Janus Au-TiO2 Photocatalysts with Strong Localization of Plasmonic Near-Fields for Efficient Visible-Light Hydrogen Generation. *Advanced Materials*, 2012. **24**(17): pp. 2310–2314.
170. Tang, Y.X., Z.L. Jiang, J.Y. Deng, D.G. Gong, Y.K. Lai, H.T. Tay, I.T.K. Joo, T.H. Lau, Z.L. Dong, and Z. Chen, Synthesis of Nanostructured Silver/Silver Halides on Titanate Surfaces and Their Visible-Light Photocatalytic Performance. *Acs Applied Materials & Interfaces*, 2012. **4**(1): pp. 438–446.
171. Tang, Y.X., Z.L. Jiang, Q.L. Tay, J.Y. Deng, Y.K. Lai, D.G. Gong, Z.L. Dong, and Z. Chen, Visible-light plasmonic photocatalyst anchored on titanate nanotubes: a novel nanohybrid with synergistic effects of adsorption and degradation. *Rsc Advances*, 2012. **2**(25): pp. 9406–9414.
172. Bian, H.D., X. Shu, J.F. Zhang, B. Yuan, Y. Wang, L.J. Liu, G.Q. Xu, Z. Chen, and Y.C. Wu, Uniformly Dispersed and Controllable Ligand-Free Silver-Nanoparticle-Decorated TiO2 Nanotube Arrays with Enhanced Photoelectrochemical Behaviors. *Chemistry-an Asian Journal*, 2013. **8**(11): pp. 2746–2754.
173. Tang, Y.X., Z.L. Jiang, G.C. Xing, A.R. Li, P.D. Kanhere, Y.Y. Zhang, T.C. Sum, S.Z. Li, X.D. Chen, Z.L. Dong, and Z. Chen, Efficient Ag@AgCl Cubic Cage Photocatalysts Profit from Ultrafast Plasmon-Induced Electron Transfer Processes. *Advanced Functional Materials*, 2013. **23**(23): pp. 2932–2940.
174. Chen, Z.Y., L. Fang, W. Dong, F.G. Zheng, M.R. Shen, and J.L. Wang, Inverse opal structured Ag/TiO2 plasmonic photocatalyst prepared by pulsed current deposition and its enhanced visible light photocatalytic activity. *Journal of Materials Chemistry A*, 2014. **2**(3): pp. 824–832.
175. Lang, X.J., X.D. Chen, and J.C. Zhao, Heterogeneous visible light photocatalysis for selective organic transformations. *Chemical Society Reviews*, 2014. **43**(1): pp. 473–486.
176. Ngaw, C.K., Q.C. Xu, T.T.Y. Tan, P. Hu, S.W. Cao, and J.S.C. Loo, A strategy for in-situ synthesis of well-defined core-shell Au@TiO2 hollow spheres for enhanced photocatalytic hydrogen evolution. *Chemical Engineering Journal*, 2014. **257**: pp. 112–121.
177. Sun, S., H.Z. Liu, L. Wu, C.E. Png, and P. Bai, Interference-Induced Broadband Absorption Enhancement for Plasmonic-Metal@Semiconductor Microsphere as Visible Light Photocatalyst. *Acs Catalysis*, 2014. **4**(12): pp. 4269–4276.
178. Xi, B.J., X.N. Chu, J.Y. Hu, C.S. Bhatia, A.J. Danner, and H. Yang, Preparation of Ag/TiO2/SiO2 films via photo-assisted deposition and adsorptive self-assembly for catalytic bactericidal application. *Applied Surface Science*, 2014. **311**: pp. 582–592.
179. Zhan, Z.Y., J.N. An, H.C. Zhang, R.V. Hansen, and L.X. Zheng, Three-Dimensional Plasmonic Photoanodes Based on Au-Embedded TiO2 Structures for Enhanced Visible-Light Water Splitting. *Acs Applied Materials & Interfaces*, 2014. **6**(2): pp. 1139–1144.

180. Zhang, X.L., J.L. Zhao, S.G. Wang, H.T. Dai, and X.W. Sun, Shape-dependent localized surface plasmon enhanced photocatalytic effect of ZnO nanorods decorated with Ag. *International Journal of Hydrogen Energy*, 2014. **39**(16): pp. 8238–8245.

181. Wang, X., Y. Tang, Z. Chen, and T.T. Lim, Highly stable heterostructured Ag-AgBr/TiO$_2$ composite: A bifunctional visible-light active photocatalyst for destruction of ibuprofen and bacteria. *Journal of Materials Chemistry*, 2012. **22**(43): pp. 23149–23158.

182. Wang, P., Y. Tang, Z. Dong, Z. Chen, and T.T. Lim, Ag-AgBr/TiO2/RGO nanocomposite for visible-light photocatalytic degradation of penicillin G. *Journal of Materials Chemistry A*, 2013. **1**(15): pp. 4718–4727.

183. Liu, C.H., M.H. Hong, Y. Zhou, G.X. Chen, M.M. Saw, and A.T.S. Hor, Synthesis and characterization of Ag deposited TiO2 particles by laser ablation in water. *Physica Scripta*, 2007. *T***129**: pp. 326–328.

184. Zheng, Z.F., J. Teo, X. Chen, H.W. Liu, Y. Yuan, E.R. Waclawik, Z.Y. Zhong, and H.Y. Zhu, Correlation of the Catalytic Activity for Oxidation Taking Place on Various TiO2 Surfaces with Surface OFF Groups and Surface Oxygen Vacancies. *Chemistry-a European Journal*, 2010. **16**(4): pp. 1202–1211.

185. Zhang, L., M.D. Han, O.K. Tan, M.S. Tse, Y.X. Wang, and C.C. Sze, Facile fabrication of Ag/C-TiO2 nanoparticles with enhanced visible light photocatalytic activity for disinfection of Escherichia coli and Enterococcus faecalis. *Journal of Materials Chemistry B*, 2013. **1**(4): pp. 564–570.

186. Zhang, Z.Y., Z. Wang, S.W. Cao, and C. Xue, Au/Pt Nanoparticle-Decorated TiO2 Nanofibers with Plasmon-Enhanced Photocatalytic Activities for Solar-to-Fuel Conversion. *Journal of Physical Chemistry C*, 2013. **117**(49): pp. 25939–25947.

187. Zhang, Z.Y., S.W. Cao, Y.S. Liao, and C. Xue, Selective photocatalytic decomposition of formic acid over AuPd nanoparticle-decorated TiO2 nanofibers toward high-yield hydrogen production. *Appl. Catal. B-Environ*, 2015. **162**: pp. 204–209.

188. Wei, Y.F., J.H. Kong, L.P. Yang, L. Ke, H.R. Tan, H. Liu, Y.Z. Huang, X.W. Sun, X.H. Lu, and H.J. Du, Polydopamine-assisted decoration of ZnO nanorods with Ag nanoparticles: an improved photoelectrochemical anode. *Journal of Materials Chemistry A*, 2013. **1**(16): pp. 5045–5052.

189. Yang, S.K., M.Y. Li, X. Zhu, G.Q. Xu, and J.H. Wu, Photochemical Synthesis of Hierarchical Multiple-Growth-Hillock Superstructures of Silver Nanoparticles on ZnO. *Journal of Physical Chemistry C*, 2015. **119**(25): pp. 14312–14318.

190. Wang, P., B. Huang, X. Zhang, X. Qin, H. Jin, Y. Dai, Z. Wang, J. Wei, J. Zhan, S. Wang, J. Wang, and M.H. Whangbo, Highly efficient visible-light plasmonic photocatalyst Ag@AgBr. *Chemistry - A European Journal*, 2009. **15**(8): pp. 1821–1824.

191. Wang, P., B.B. Huang, Q.Q. Zhang, X.Y. Zhang, X.Y. Qin, Y. Dai, J. Zhan, J.X. Yu, H.X. Liu, and Z.Z. Lou, Highly Efficient Visible Light Plasmonic Photocatalyst Ag@Ag(Br,I). *Chemistry-a European Journal*, 2010. **16**(33): pp. 10042–10047.

192. Cao, S.W., J. Fang, M.M. Shahjamali, F.Y.C. Boey, J. Barber, S.C.J. Loo, and C. Xue, Plasmon-Enhanced Hydrogen Evolution on Au-InVO4 Hybrid Microspheres. *Rsc Advances*, 2012. **2**(13): pp. 5513–5515.

193. Wang, G.N., P. Wang, H.K. Luo, and T.S.A. Hor, Novel Au/La-SrTiO3 microspheres: Superimposed Effect of Gold Nanoparticles and Lanthanum Doping in Photocatalysis. *Chemistry-an Asian Journal*, 2014. **9**(7): pp. 1854–1859.
194. Sun, S., W. Wang, L. Zhang, M. Shang, and L. Wang, Ag@C core/shell nanocomposite as a highly efficient plasmonic photocatalyst. *Catalysis Communications*, 2009. **11**(4): pp. 290–293.
195. Zhang, H., X. Fan, X. Quan, S. Chen, and H. Yu, Graphene sheets grafted Ag@AgCl hybrid with enhanced plasmonic photocatalytic activity under visible light. *Environmental Science and Technology*, 2011. **45**(13): pp. 5731–5736.
196. Zhu, M., P. Chen, and M. Liu, Graphene oxide enwrapped Ag/AgX (X = Br, Cl) nanocomposite as a highly efficient visible-light plasmonic photocatalyst. *ACS Nano*, 2011. **5**(6): pp. 4529–4536.
197. Chen, X., H.Y. Zhu, J.C. Zhao, Z.F. Zheng, and X.P. Gao, Visible-light-driven oxidation of organic contaminants in air with gold nanoparticle catalysts on oxide supports. *Angewandte Chemie - International Edition*, 2008. **47**(29): pp. 5353–5356.
198. Foster, H.A., I.B. Ditta, S. Varghese, and A. Steele, Photocatalytic disinfection using titanium dioxide: spectrum and mechanism of antimicrobial activity. *Applied Microbiology and Biotechnology*, 2011. **90**(6): pp. 1847–1868.
199. Sun, D.D., J.H. Tay, and K.M. Tan, Photocatalytic degradation of E-coliform in water. *Water Res*, 2003. **37**(14): pp. 3452–3462.
200. Pal, A., S.O. Pehkonen, L.E. Yu, and M.B. Ray, Photocatalytic Inactivation of Airborne Bacteria in a Continuous-Flow Reactor. *Ind. Eng. Chem. Res*, 2008. **47**(20): pp. 7580–7585.
201. Ng, J.W., X.W. Zhang, T. Zhang, J.H. Pan, J.H.A. Du, and D.D. Sun, Construction of self-organized free-standing TiO2 nanotube arrays for effective disinfection of drinking water. *J. Chem. Technol. Biotechnol*, 2010. **85**(8): pp. 1061–1066.
202. Ma, J.Z., Z.G. Xiong, T.D. Waite, W.J. Ng, and X.S. Zhao, Enhanced inactivation of bacteria with silver-modified mesoporous TiO2 under weak ultraviolet: irradiation. *Microporous and Mesoporous Materials*, 2011. **144**(1–3): pp. 97–104.
203. Liu, L., Z.Y. Liu, H.W. Bai, and D.D. Sun, Concurrent filtration and solar photocatalytic disinfection/degradation using high-performance Ag/TiO2 nanofiber membrane. *Water Res*, 2012. **46**(4): pp. 1101–1112.
204. Bai, H.W., Z.Y. Liu, L. Liu, and D.D. Sun, Large-Scale Production of Hierarchical TiO2 Nanorod Spheres for Photocatalytic Elimination of Contaminants and Killing Bacteria. *Chemistry-a European Journal*, 2013. **19**(9): pp. 3061–3070.
205. Xiong, P., and J.Y. Hu, Inactivation/reactivation of antibiotic-resistant bacteria by a novel UVA/LED/TiO2 system. *Water Res*, 2013. **47**(13): pp. 4547–4555.
206. Sun, D.D., Y. Wu, and P. Gao, Effects of TiO2 nanostructure and operating parameters on optimized water disinfection processes: A comparative study. *Chemical Engineering Journal*, 2014. **249**: pp. 160–166.
207. Sunada, K., T. Watanabe, and K. Hashimoto, Studies on photokilling of bacteria on TiO_2 thin film. *Journal of Photochemistry and Photobiology A: Chemistry*, 2003. **156**(1–3): pp. 227–233.

208. Cho, M., H. Chung, W. Choi, and J. Yoon, Linear correlation between inactivation of *E. coli* and OH radical concentration in TiO_2 photocatalytic disinfection. *Water Res,*, 2004. **38**(4): pp. 1069–1077.
209. Blake, D.M., P.C. Maness, Z. Huang, E.J. Wolfrum, J. Huang, and W.A. Jacoby, Application of the photocatalytic chemistry of titanium dioxide to disinfection and the killing of cancer cells. *Separation and Purification Methods*, 1999. **28**(1): pp. 1–50.
210. Byrne, J.A., P.A. Fernandez-Ibañez, P.S.M. Dunlop, D.M.A. Alrousan, and J.W.J. Hamilton, Photocatalytic enhancement for solar disinfection of water: A review. *International Journal of Photoenergy* 2011, **2011**.
211. Rincón, A.G., and C. Pulgarin, Field solar *E. coli* inactivation in the absence and presence of TiO_2: Is UV solar dose an appropriate parameter for standardization of water solar disinfection? *Solar Energy*, 2004. **77**(5): pp. 635–648.
212. Chin, S.S., K. Chiang, and A.G. Fane, The stability of polymeric membranes in a TiO2 photocatalysis process. *Journal of Membrane Science*, 2006. **275**(1–2): pp. 202–211.
213. Goei, R., T.-T. Lim, Asymmetric TiO_2 hybrid photocatalytic ceramic membrane with porosity gradient: Effect of structure directing agent on the resulting membranes architecture and performances. *Ceramics International*, 2014. **40**(5): pp. 6747–6757.
214. Goei, R., and T.T. Lim, Ag-decorated TiO_2 photocatalytic membrane with hierarchical architecture: Photocatalytic and anti-bacterial activities. *Water Res*, 2014. **59**: pp. 207–218.
215. Pignatello, J.J., E. Oliveros, and A. MacKay, Advanced Oxidation Processes for Organic Contaminant Destruction Based on the Fenton Reaction and Related Chemistry. *Crit. Rev. Environ. Sci. Technol*, 2006. **36**(1): pp. 1–84.
216. Hu, Z.T., B. Chen, and T.T. Lim, Single-crystalline Bi2Fe4O9 synthesized by low-temperature co-precipitation: Performance as photo- and Fenton catalysts. *RSC Advances*, 2014. **4**(53): pp. 27820–27829.
217. Oh, W.D., S.K. Lua, Z. Dong, and T.T. Lim, High surface area DPA-hematite for efficient detoxification of bisphenol A via peroxymonosulfate activation. *Journal of Materials Chemistry A*, 2014. **2**(38): pp. 15836–15845.
218. Oh, W.D., S.K. Lua, Z. Dong, and T.T. Lim, A novel three-dimensional spherical $CuBi_2O_4$ nanocolumn arrays with persulfate and peroxymonosulfate activation functionalities for 1H-benzotriazole removal. *Nanoscale* 2015.
219. Goh, K.H., T.T. Lim, and Z. Dong, Application of layered double hydroxides for removal of oxyanions: A review. *Water Res*, 2008. **42**(6–7): pp. 1343–1368.
220. Goh, K.H. Sorption of Oxyanions on Nanocrystalline Mg/Al Layered Double Hydroxides: Sorption Characteristics, Mechanisms, and Matrix Interferences. Nanyang Technological University, Singapore, 2010.
221. Goh, K.H., T.T. Lim, A. Banas, and Z. Dong, Sorption characteristics and mechanisms of oxyanions and oxyhalides having different molecular properties on Mg/Al layered double hydroxide nanoparticles. *Journal of Hazardous Materials*, 2010. **179**(1–3): pp. 818–827.
222. Goh, K.H., T.T. Lim, and Z. Dong, Enhanced arsenic removal by hydrothermally treated nanocrystalline MG/AL layered double hydroxide with nitrate intercalation. *Environmental Science and Technology*, 2009. **43**(7): pp. 2537–2543.

223. Tang, Y.X., Y.K. Lai, D.G. Gong, K.H. Goh, T.T. Lim, Z.L. Dong, and Z. Chen, Ultrafast Synthesis of Layered Titanate Microspherulite Particles by Electrochemical Spark Discharge Spallation. *Chemistry-a European Journal*, 2010. **16**(26): pp. 7704–7708.
224. Tang, Y.X., D.G. Gong, Y.K. Lai, Y.Q. Shen, Y.Y. Zhang, Y.Z. Huang, J. Tao, C.J. Lin, Z.L. Dong, and Z. Chen, Hierarchical layered titanate microspherulite: formation by electrochemical spark discharge spallation and application in aqueous pollutant treatment. *Journal of Materials Chemistry*, 2010. **20**(45): pp. 10169–10178.
225. Peric, J., M. Trgo, and N.V. Medvidovic, Removal of zinc, copper and lead by natural zeolite — a comparison of adsorption isotherms. *Water Res*, 2004. **38**(7): pp. 1893–1899.
226. Zhuang, Y., Y. Yang, G.L. Xiang, and X. Wang, Magnesium Silicate Hollow Nanostructures as Highly Efficient Absorbents for Toxic Metal Ions. *Journal of Physical Chemistry C*, 2009. **113**(24): pp. 10441–10445.
227. Sreejalekshmi, K., K.A. Krishnan, and T. Anirudhan, Adsorption of Pb (II) and Pb (II)-citric acid on sawdust activated carbon: Kinetic and equilibrium isotherm studies. *Journal of Hazardous Materials*, 2009. **161**(2): pp. 1506–1513.
228. Singh, C., J. Sahu, K. Mahalik, C. Mohanty, B.R. Mohan, and B. Meikap, Studies on the removal of Pb (II) from wastewater by activated carbon developed from Tamarind wood activated with sulphuric acid. *Journal of Hazardous Materials*, 2008. **153**(1): pp. 221–228.
229. Oubagaranadin, J.U.K., and Z.V.P. Murthy, Adsorption of Divalent Lead on a Montmorillonite-Illite Type of Clay. *Ind. Eng. Chem. Res*, 2009. **48**(23): pp. 10627–10636.
230. Entezari, M.H., and T.R. Bastami, Sono-sorption as a new method for the removal of lead ion from aqueous solution. *Journal of Hazardous Materials*, 2006. **137**(2): pp. 959–964.
231. Yang, D.J., Z.F. Zheng, H.W. Liu, H.Y. Zhu, X.B. Ke, Y. Xu, D. Wu, and Y. Sun, Layered Titanate Nanofibers as Efficient Adsorbents for Removal of Toxic Radioactive and Heavy Metal Ions from Water. *Journal of Physical Chemistry C*, 2008. **112**(42): pp. 16275–16280.
232. Lim, Y.W.L., Y.X. Tang, Y.H. Cheng, and Z. Chen, Morphology, crystal structure and adsorption performance of hydrothermally synthesized titania and titanate nanostructures. *Nanoscale*, 2010. **2**(12): pp. 2751–2757.
233. Lim, S.F., Y.M. Zheng, and J.P. Chen, Organic arsenic adsorption onto a magnetic sorbent. *Langmuir*, 2009. **25**(9): pp. 4973–4978.
234. Ma, Y., Y.M. Zheng, and J.P. Chen, A zirconium based nanoparticle for significantly enhanced adsorption of arsenate: Synthesis, characterization and performance. *Journal of Colloid and Interface Science*, 2011. **354**(2): pp. 785–792.
235. Wei, Y.T., Y.M. Zheng, and J.P. Chen, Uptake of methylated arsenic by a polymeric adsorbent: Process performance and adsorption chemistry. *Water Res*, 2011. **45**(6): pp. 2290–2296.

236. Zhang, G., A. Khorshed, and J. Paul Chen, Simultaneous removal of arsenate and arsenite by a nanostructured zirconium-manganese binary hydrous oxide: Behavior and mechanism. *Journal of Colloid and Interface Science*, 2013. **397**: pp. 137–143.
237. Yu, Y., and J.P. Chen, Fabrication and performance of a Mn-La metal composite for remarkable decontamination of fluoride. *Journal of Materials Chemistry A*, 2014. **2**(21): pp. 8086–8093.
238. Chen, C., P. Gunawan, and R. Xu, Self-assembled Fe3O4-layered double hydroxide colloidal nanohybrids with excellent performance for treatment of organic dyes in water. *Journal of Materials Chemistry*, 2011. **21**(4): pp. 1218–1225.
239. Chen, C., P. Wang, T.T. Lim, L. Liu, S. Liu, and R. Xu, A facile synthesis of monodispersed hierarchical layered double hydroxide on silica spheres for efficient removal of pharmaceuticals from water. *Journal of Materials Chemistry A*, 2013. **1**(12): pp. 3877–3880.
240. Song, X., Y. Wang, K. Wang, and R. Xu, Low-cost carbon nanospheres for efficient removal of organic dyes from aqueous solutions. *Industrial and Engineering Chemistry Research*, 2012. **51**(41): pp. 13438–13444.
241. Song, X., P. Gunawan, R. Jiang, S.S.J. Leong, K. Wang, and R. Xu, Surface activated carbon nanospheres for fast adsorption of silver ions from aqueous solutions. *Journal of Hazardous Materials*, 2011. **194**: pp. 162–168.
242. Wang, B., H. Wu, L. Yu, R. Xu, T.T. Lim, and X.W. Lou, Template-free formation of uniform urchin-like α-FeOOH hollow spheres with superior capability for water treatment. *Advanced Materials*, 2012. **24**(8): pp. 1111–1116.
243. Yap, P.S., and T.T. Lim, Solar regeneration of powdered activated carbon impregnated with visible-light responsive photocatalyst: Factors affecting performances and predictive model. *Water Res*, 2012. **46**(9): pp. 3054–3064.
244. Oh, W.D., S.K. Lua, Z. Dong, and T.T. Lim, Performance of magnetic activated carbon composite as peroxymonosulfate activator and regenerable adsorbent via sulfate radical-mediated oxidation processes. *Journal of Hazardous Materials*, 2015. **284**: pp. 1–9.
245. Bhattacharyya, A., S. Kawi, and M.B. Ray, Photocatalytic degradation of orange II by TiO2 catalysts supported on adsorbents. *Catal. Today*, 2004. **98**(3): pp. 431–439.

5. Solar Energy and Energy Storage Materials and Devices Research in Singapore

D. Sabba*, J. Wang[†], M. Srinivasan[‡],
A.G. Aberle[§] and S. Mhaisalkar*,[‡],[¶]

"No challenge poses more of a public threat than climate change," President of the United States of America, Barack Obama, 11 April 2015

"In the material sciences these are and have been, and are most surely likely to continue to be heroic days." J. Robert Oppenheimer, 1904–1967

World population increase, urbanization, rising living standards, climate change, and the need for secure and safe low-carbon energy sources constitute the key global megatrends of today; and represent an unparalleled threat and opportunity to business-as-usual.

For the sustenance of the 9.6 billion people on the planet by 2050 (up from 7.2 billion today) and to meet their expectations for a high standard of living, we will need to account for the worldwide primary energy consumption rising to 210 TWh/year by 2050; compared with 155 TWh/year in 2015.

It is imperative that we address this energy demand in the context of the anthropogenic CO_2 that affects the global average surface temperature. Under business-as-usual scenarios, the atmospheric CO_2 concentration in 2050 is expected to reach 685 ppm; compared with 400 ppm in 2015. At this CO_2 level, the global average temperature will rise by 3°C and the effects it will have on food, water and the environment will be devastating. In order to minimise environmental impact, scientists have converged on a consensus for

---

*Energy Research Institute @ NTU (ERI@N), Singapore.
[†]Department of Materials Science and Engineering, National University of Singapore.
[‡]School of Materials Science and Engineering, Nanyang Technological University.
[§]Solar Energy Research Institute of Singapore (SERIS), National University of Singapore.
[¶]Subodh@ntu.edu.sg.

a need for action on a "2DS": 2°C Scenario, that will limit the CO_2 level to 450 ppm (by 2035) and a global average temperature rise to 2°C. Any action to limit CO_2 will need massively scaling renewable energy development and installation, and will need breakthroughs in all aspects of science, technology, and business.

Solar photovoltaics ('PV') have by far the greatest potential for renewable energy generation, with the total solar energy received by Earth's oceans and land masses being approximately 1 million TWh/yr. There is more solar energy available in one hour than the amount mankind uses in one year. In contrast, the total potential for wind, marine and other renewables put together is less than 0.4% compared with that of solar. The installation cost of solar photovoltaics has dropped enormously from US$10/$W_p$ ('watt-peak') in 2000 to close to $1/$W_p$ in 2015, and is now at, or even below, grid parity in most countries worldwide. Massive scaling of solar PV will need (i) increased efficiency, (ii) reduced costs, (iii) new technologies to explore alternate options such as building-integrated photovoltaics (BIPV).

The intermittency of electricity generation is the most severe concern with all renewables, and is even more significant with PV which become inactive during the dark hours of the day. Energy storage is essential to mitigate the impact of all renewables and currently represents the weak link in the deployment of renewables. While the levelized cost of electricity (LCOE) for PV has dropped to under US$0.10/kWh, the cost-adder (to LCOE) of energy storage is estimated in the range of US$0.20 to 0.30/kWh. For applications in electric vehicles and for integration with renewables, the capital cost of energy storage needs to drop from close to US$500/kWh to under US$200/kWh for Li ion batteries. Beyond costs, the key concerns with energy storage include cycle life, reliability, and safety.

To summarize the discussions above, the key opportunity in the curtailment of the CO_2 increase in the 2DS would include a massive emphasis on deploying renewable energies. In particular solar — being the largest available resource — offers the greatest potential. The factors that would accelerate the deployment of photovoltaics include increasing efficiency, reducing costs, and finally developing low cost, reliable energy storage technologies to mitigate the effect of intermittency on the electricity grid. It is in this context that nanomaterials could offer breakthrough solutions and could play a transformative role in the generation and storage of solar energy.

1. PV Materials and Devices Research in Singapore

1.1 Introduction

There was little research done on PV materials and devices in Singapore prior to 2006. However, the situation changed completely in 2007 when the government announced a major funding initiative in the cleantech sector. In the PV space, noteworthy results of this SGD 700 million initiative included the establishment of the

Solar Energy Research Institute of Singapore (SERIS) at the National University (NUS) of Singapore (in April 2008), and the construction in Tuas, in western Singapore, of one of the world's largest factories for crystalline silicon wafers, solar cells, and PV modules by the Norwegian company Renewable Energy Corporation ASA ("REC"). REC's US$ 2 billion factory was officially opened in 2010 by Singapore's Prime Minister, with an annual production capacity of 800 Megawatt-peak (MW_p) for wafers, cells, and PV modules. PV research activities also increased at NTU (in particular at the Energy Research Institute @NTU, ERI@N), at NUS (Faculty of Engineering and Faculty of Science), and at A*STAR.

1.2. Silicon wafer solar cells

The most common solar cells are p-n junction silicon cells. Upon doping with boron and phosphorus atoms, silicon becomes a "p" and "n" type semiconductor respectively, and when these two materials are stacked together they form a p-n junction, which features a strong built-in electric field. Upon illuminating the p-n junction, photons excite electrons from the valence band to the conduction band, thereby generating electron-hole pairs. The electron-hole pairs are separated by the electric field of the p-n junction, leading to the build-up of negative charge on the n-type silicon surface and positive charge on the p-type silicon surface. Metal electrodes on both surfaces of the solar cell enable the extraction of the light-generated electric power. Crystalline silicon wafer solar cells have been dominating the global PV market since the 1980s. As such it was a natural choice in 2008 to make this PV technology the backbone of the R&D activities of the newly formed Solar Energy Research Institute of Singapore (SERIS). This proved to be a wise decision, given the fact that silicon wafer based solar cells have further increased their global PV market share to over 90% in 2015, while at the same time dramatically reducing the manufacturing costs of the resulting PV modules to below 0.5 US$/$W_p$.

As shown in Table 1, the commercially available silicon wafer solar cells can broadly be categorised into four groups, ranging from standard p-type homojunction cells to heterojunction cells on n-type wafers. SERIS has established strong R&D capabilities for each of the four groups of industrially important silicon wafer solar cells. At its main location on the NUS campus, the institute built a 1200 m^2 silicon solar cell laboratory hosting R&D pilot lines for screen-printed multicrystalline and monocrystalline silicon wafer solar cells. The pilot lines are able to process full-size wafers (156 mm x 156 mm) with a production rate of several hundred wafers per hour. Furthermore, SERIS is presently setting up an R&D pilot line for heterojunction silicon wafer solar cells. The lab at SERIS is one the largest and most advanced laboratories for industrial silicon wafer solar cells at any public research organisation in the world.

Table 1. Overview of the most important industrial silicon wafer based solar cell technologies (Status: June 2015).

Solar Cell Technology	Present Cell Efficiencies in Industry	Remarks
Standard p-type homo-junction cells ('Al-BSF' cells)	16.5–18% on multi-Si wafers, 18–19% on mono-Si wafers	Screen-printed and co-fired contacts, dominant Si wafer cell (> 80 % of all Si wafer cells are of this type), many manufacturers
Advanced p-type homo-junction cells	18–19% on multi-Si wafers, 19–20.5% on mono-Si wafers	Screen-printed and co-fired contacts, dielectrically passivated rear surface, > 10 manufacturers (Trina Solar, Hanwha Q Cells, REC Solar, Jinko Solar, SolarWorld, JA Solar, Suntech, Sunrise, ...)
N-type homojunction cells on mono-Si wafers	19–22% for standard cells, 22–25% for all-back-contact cells	Screen-printed or plated contacts, few manufacturers (Yingli, Sharp, Neo Solar Power, Sunrise, SunPower, ...)
N-type heterojunction cells on mono-Si wafers	20–22.5%	Bifacial cells, heterojunction via a-Si:H, screen-printed or plated contacts, very few manufacturers (Panasonic, ...)

Over the past few years, the main research topics on silicon wafer solar cells at SERIS included:

- Improved wet-chemical processing methods
- Improved junction formation methods
- Improved surface passivation methods
- Improved metallisation methods
- Advanced simulation and characterisation methods

For example, SERIS conducted a comprehensive study on the benefits and potential of inline phosphorus diffusion for silicon solar cells. Inline diffusion, which uses conventional belt furnaces with phosphoric acid (H_3PO_4) as phosphorus source, has the potential of being a lower-cost and higher-throughput alternative to the classical tube diffusion process. By having the dopant source directly deposited onto the wafer surface, this typically achieves a better doping homogeneity compared to tube diffusion, especially for emitters with a high sheet resistance (R_{sq}). However, the use of an open-ended furnace, lower-grade chemical precursors, metal conveyor belts and shorter process times have their own adverse impacts on the solar cell efficiency. Additionally, a surface 'dead layer' forming at the diffused surface by a high dopant concentration at the surface, as well as surface contaminants resulting from the direct deposition of dopants onto the wafers, lower the efficiencies of inline-diffused emitter (ILDE) solar cells, owing to the reduced short-circuit current (I_{sc}) and open-circuit voltage (V_{oc}).[1] Thus, the strategy for achieving high PV efficiency for IDLE solar cells lies in the thinning, or even complete removal, of the dead layer.

In 2013, SERIS introduced a new non-acidic and uniform n⁺ emitter etch-back process, the SERIS etch.[2] On ILDE multi-Si solar cells, the SERIS etch can be used for emitter etch-backs of up to 30 Ω/square, maintaining front surface morphology and R_{sq} uniformity. The effect of the back etch on the Al-BSF multi-Si solar cell efficiency is shown in Figure 1(b), where four groups of solar cells were fabricated, all featuring a 70 Ω/sq emitter. Group 1 had an as-diffused emitter, whereas the emitters of Groups 2 to 4 were obtained by etching back emitters with as-diffused sheet resistances of 60, 50 and 40 Ω/sq, respectively. Due to the etch-back, Group 4 had the lowest surface concentration but a larger junction depth compared to the other groups which is beneficial for the metallisation process, as it reduces the possibility of shunting. Groups 3 and 4 achieved median solar cell efficiencies of about 17.7%, representing a 0.4% (absolute) efficiency gain compared to the median efficiency of the cells without an etch-back (group 1). These results show that the etch-back process utilising the SERIS etch can enable higher efficiencies for industrial multi-Si wafer solar cells.

Another example for SERIS' silicon solar cell research is Griddler (Figure 2), a finite element analyser for complex solar cell metallisation patterns to accurately

(a)

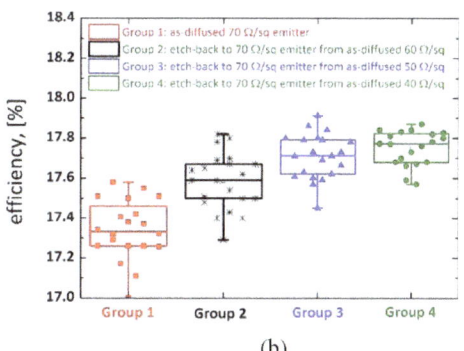
(b)

Fig. 1. (a) Photograph of REC's manufacturing facility for silicon wafers, cells and modules in Tuas, Singapore. In 2015 the production capacity for PV modules was increased to 1.3 GW_p per year. Photo courtesy of REC Solar. (b) Distribution of the 1-Sun efficiencies of Al-BSF multi-Si solar cells with the as-diffused n⁺ emitter (Group 1: as-diffused 70 W/sq) and the etched-back n⁺ emitters (Groups 2–4: etch-back to 70 W/sq).

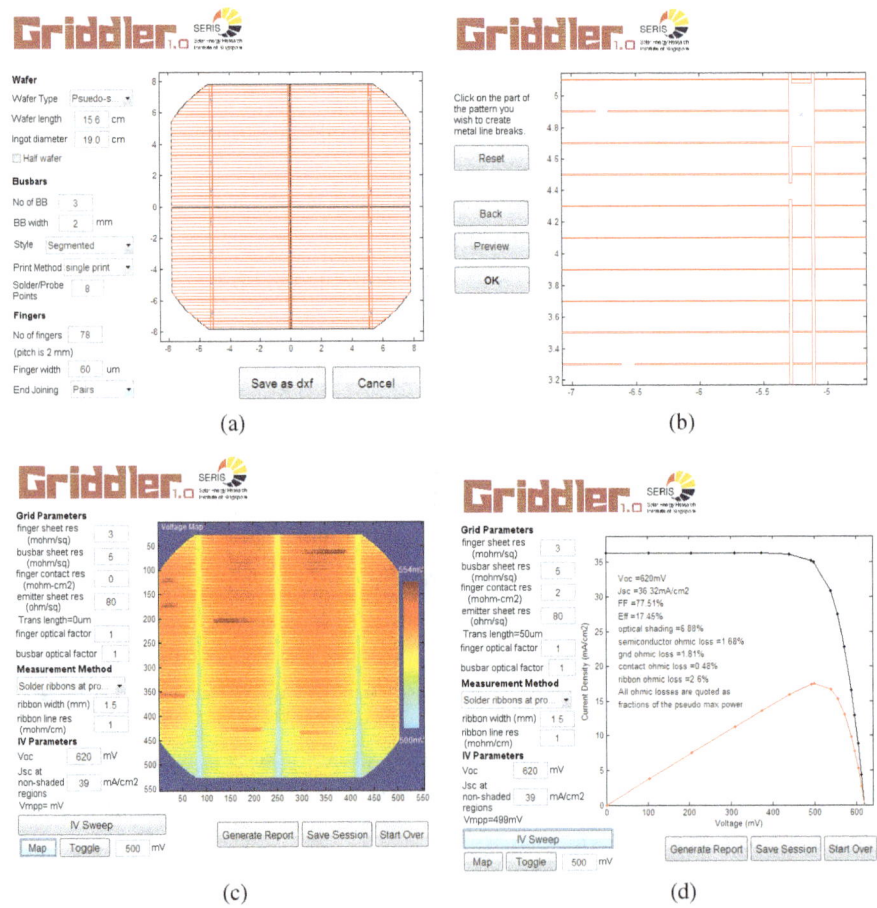

Fig. 2 Screen shots from Griddler 1.0. (a) Interface to define the solar cell front grid pattern; (b) User may introduce line breaks into the pattern to assess the level of impact on the device performance; (c) Simulated voltage distribution; (d) Simulated I-V curve.

simulate solar cells with complex metallisation geometries, such as metal-wrap-through (MWT) cells, interdigitated-back-contact (IBC) cells, and metal grids with non-ideal features like finger breaks, finger striations and non-uniform contact resistance. SERIS developed a programme that captures the cell metallisation geometry from CAD files, and efficiently meshes the cell plane for finite element analysis, yielding standard data such as the I-V curve, voltage and R_s distribution. The programme also features a powerful post processor that predicts the rate of change in efficiency with respect to the incremental changes in the metallization pattern, opening up the possibility of intelligent computer aided design procedures.

The simplest and most common way to analyse the series resistance R_s in solar cells involves solving the ohmic power loss in each current carrying element (semiconductor, fingers, busbars) in suitably defined unit cells using approximations in the current flow pattern as well as the cell voltage. While this approach is easy to implement and accurate where the R_s is small, it becomes inadequate in the following situations: (1) The cell design contains significant areas that cannot be represented by repeating unit cells, such as those found in many MWT, IBC and concentrator solar cells; (2) The designer wishes to study the impact of non-ideal features like finger breaks, finger striations, and non-uniform contact resistance; (3) Rather than just R_s, the designer seeks detailed outputs such as the cell voltage distribution and current-voltage (I-V) characteristics for comparisons to measured data from luminescence imaging, I-V measurements, etc. In order to help tackle the general metallisation pattern problem on a rigorous basis, SERIS has developed Griddler, a finite-element-method (FEM) based programme utilising sparse matrix methods.[3] Since 2015 SERIS has been offering Griddler 2.0 as a stand-alone product.[4] In addition, the same team behind the computer programme also carries much hands-on experience with real-world cell metallisation scheme design and processing. Based on these practical and analytical abilities, SERIS offers consultancy services in the area of solar cell metallisation scheme design, with respect to constraints such as the metal paste used, line widths, amount of silver and sheet resistance, to deliver designs that are optimised on paper and robust in production.

1.3. Silicon thin-film solar cells

Due to a greatly reduced semiconductor material consumption and the ability to (i) fabricate the solar cells on inexpensive large-area foreign substrates and (ii) to monolithically series-connect the fabricated solar cells, thin-film PV have the potential of achieving the lowest module costs ($/$W_p$) of all PV technologies.

Although amorphous silicon (a-Si:H) deposited by plasma-enhanced chemical vapour deposition (PECVD) at about 200°C, which is the baseline thin-film PV technology[4], possesses a number of excellent properties for low-cost PV electricity — including a high optical absorption coefficient (enabling very thin absorber thicknesses of 300 nm or less), large-area silicon diode deposition at low temperature (~200°C) onto rigid or flexible substrates, and monolithic series interconnection of the individual cells — it was not able to conquer a significant share of the global PV market owing to the low stable average efficiency of 6% or less of large-area single-junction PV modules.[5] One factor behind this modest stable efficiency is the "Staebler–Wronski effect"[6], i.e. the light-induced degradation of the initial module efficiency to the stabilized module efficiency. Research is continuing into finding ways to reduce this effect, however, after two decades of intense global research the prospects seem limited.

By replacing a-Si:H with polycrystalline silicon ("pc-Si") and by exploiting pc-Si's superior electronic properties (carrier lifetimes and mobilities, minority carrier diffusion length, doping efficiency), efficiencies of > 10% have been achieved. This approach was pioneered by the Japanese company Sanyo Electric which showed in the early 1990s that solid phase crystallisation (SPC) at ~600°C of a relatively thick (~5 µm) PECVD-deposited a-Si film gives 9.2% efficiency on a metal substrate.[7] The cells featured a p^+ doped a-Si heterojunction emitter on an n^+n polycrystalline silicon structure crystallised by SPC. In the late 1990s, Pacific Solar Pty Ltd, a spin-off company of the University of New South Wales (UNSW) in Australia, successfully transferred the PECVD-based SPC approach to borosilicate glass sheets (Borofloat33 from Schott AG, Germany). The best efficiency obtained so far with this so-called CSG (Crystalline Silicon on Glass) technology is 10.4%, realised in 2007 with a 94-cm², 20-cell mini-module with a fill factor (FF) of 72.1%, a J_{sc} of 29.5 mA/cm², and an average cell V_{oc} of 492 mV[8]. Independent research on PECVD-deposited n^+pp^+ SPC pc-Si solar cells on glass ("PLASMA" cells) in A.G. Aberle's previous group at UNSW had led to cell efficiencies of 9%. The best cells had a V_{oc} of around 500 mV, a FF of around 70%, and a J_{sc} of around 26 mA/cm². Cell area was 4 cm² and the silicon film thickness was in the 2–3 µm range.

As part of SERIS' ramp-up activities in 2008–2011, Aberle and co-workers established an R&D pilot line on the NUS campus for polycrystalline silicon thin-film solar cells on glass, to further develop the PLASMA solar cell technology and ramp it up to pilot-scale glass substrate sizes of 300 mm x 400 mm. Silicon deposition was conducted in a PECVD cluster tool, using the parallel-plate method and the standard industrial plasma excitation frequency (13.56 MHz). The main R&D tasks were the upscaling of all processes to the large substrate size, the development of a faster PECVD deposition process for a-Si:H, and the development of superior post-deposition processes (rapid thermal anneal, hydrogenation) to boost the pc-Si material quality.

Major results obtained at SERIS included:

- Achievement of Lambertian limited light trapping over a large area (300 mm × 400 mm) using aluminium-induced texture (AIT) glass
- Development of a phenomenological model of the AIT glass texturing process
- Development of a novel pc-Si rear surface plasma texturing etch to boost the light trapping performance
- Realisation of 24 mA/cm² in short-circuit current for a pc-Si solar cell of only 1.6 µm thickness deposited on AIT glass using a SiO_2 point contact back reflector

The other possible path towards higher efficiency for silicon thin-film PV modules is to combine the amorphous silicon cell (1.7 eV) with a second cell having a lower bandgap (1 eV), giving a double-junction (or "tandem" or "micromorph") solar cell for enhanced light harvesting. This silicon device was pioneered by the University of Neuchatel, Switzerland in the 1990s, using microcrystalline silicon (μc-Si:H), consisting of both amorphous and crystalline regions, formed by PECVD[10] which yielded initial efficiencies of up to 13.1% in 1994.[8] Importantly, because of the low J_{sc} of stable a-Si:H cells (~13 mA/cm²), the thickness of the μc-Si:H cell in the stack does not have to be increased significantly compared to a stand-alone μc-Si:H cell.

During 2010 to 2013 in a publicly funded project, SERIS tried to demonstrate that the efficiencies of micromorph thin-film PV cells and modules can be improved by employing a novel category of advanced superstrates based on nanostructured glass surfaces to enhance light absorption within both the amorphous and the microcrystalline silicon cell, while preserving an adequate morphology for the growth of the layers. With the objective of scattering the red and infrared part of solar spectrum to be absorbed by the microcrystalline cell and the blue part to be absorbed by the amorphous cell, nanostructured glass and rough transparent conductive oxide (TCO) are imperative. The project was carried out in collaboration with the Institute of Microengineering (IMT) of the Ecole Polytechnique Fédérale de Lausanne (EPFL), Switzerland. Major accomplishments of this project comprised the establishment of an R&D pilot line for micromorph silicon thin-film solar cells on glass at SERIS, capable of (i) texturing large-area (300 mm x 400 mm) glass sheets (borosilicate and soda-lime) with the aluminium induced texturization (AIT) method, see Figure 3(b); (ii) TCO deposition capability with TCO texturing; (iii) microcrystalline silicon deposition capability (undoped & doped) and solar cells; (iv) amorphous silicon deposition capability (undoped & doped) and solar cells; (v) micromorph solar cells; and (vi) amorphous and micromorph PV mini-modules. In 2012 SERIS' micromorph silicon thin-film R&D pilot line demonstrated the capability for producing 9.6% efficient (initial) amorphous solar cells with an area of

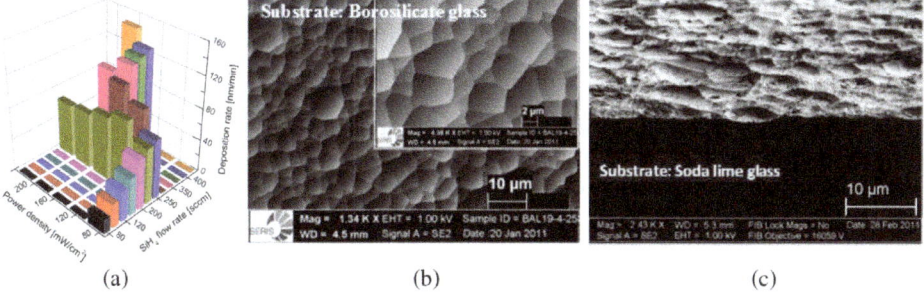

Fig. 3. (a) Deposition rate of lightly p-type doped a-Si:H films on SiN-coated planar glass substrates as a function of the plasma power and the SiH_4 flow rate. SEM images of borosilicate and soda-lime glass sheets textured with the AIT method ((b) surface view and (c) cross-sectional view).

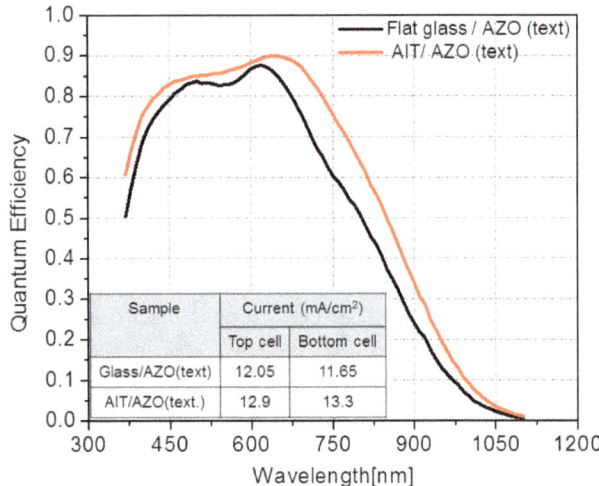

Fig. 4. Measured external quantum efficiencies of two micromorph silicon solar cells, one deposited on flat glass and the other on AIT glass. Both cells have a 900 nm thick front AZO film that was wet-chemically textured using 0.3% HCl.

1 cm², 10.1% efficient (initial) micromorph tandem cells with an area of 1 cm², and 7.7% (initial) micromorph mini-modules with an area of 104 cm². The collaboration with IMT and PVcomB (Germany) showed experimentally that the SERIS-developed advanced superstrates based on nanostructured glass surfaces (Figure 4) significantly improve the long-wavelength response of the micromorph bottom cells, as a result of improved light trapping, leading to a stable micromorph cell efficiency of 11.2% for a total Si thickness of less than 1.6 μm and a 9.85% efficiency for micromorph mini-modules with an area of 64 cm².

Given the enormous reduction of the prices of silicon wafer based PV modules in recent years, the industrial prospects of all silicon thin-film solar cell technologies look bleak at present. Major breakthroughs with the stable efficiencies of laboratory solar cells would be required before the PV industry would seriously re-consider to go into mass production with any of these PV technologies.

1.4. CIS, CIGS, CZTS family of thin-film solar cells

The Cu-chalcopyrite group of materials, CIGS, CIS, and CZTS, i.e. $Cu(In,Ga)(S,Se)_2$, $CuIn(S,Se)_2$, and $CuZnSn(S,Se)_2$, have been investigated as absorber materials for thin film solar cells because of their high absorption coefficient and suitable bandgap for solar absorption, with a record efficiency of 21.7%. One benefit of Cu chalcopyrite materials is that they perform well in low light conditions, making them suitable for applications in cloudy or shaded areas. Together with its versatility to be printed on flexible substrate, a Cu-chalcopyrite solar cell is the perfect candidate for Building Integrated PV, for self-powered smart windows, or for the roofs of pedestrian linkways in the parks. Within the framework of an NRF funded grant awarded to ERI@N and SERIS, an R&D pilot line for CIGS solar cells and mini-modules on glass substrates has been established at SERIS. The line is operational since Q3 2015. The initial goal was to develop capabilities to prepare CIGS mini-modules (area: 400 cm^2) with an efficiency of more than 12%. Another initial goal was to develop Cd-free buffer layers for a greener CIGS solar cell. The Cd-free buffer layer will be developed at ERI@N and will be integrated together with the absorber layer developed at SERIS. The mid-term goal was to develop CIGS mini-modules with efficiencies approaching 20%.

While the efforts at SERIS focused on physical vapour deposition (PVD) methodologies, the efforts at ERI@N focused on the solution processed CIGS and CZTS family of solar cells. ERI@N pioneered a fabrication method — of using the solution process to fabricate CIGS, CZTS thin film solar cells — which relies on using non-toxic chemicals and promotes easy scalability for commercial production. The efficiency of our CZTS solar cell is 9.2%, which is the highest reported for sulphide-based thin film solar cells. As a more environmental friendly approach, aqueous spray pyrolysis directly on the Mo substrate for CIS fabrication was employed and to prevent substrate oxidation, barrier layers composed of very fine grains were imperative. However the detrimental effect to charge transport by the fine grains was addressed by enlarging the grains via high temperature annealing under various vapours such as S, Se, Ar, followed by Sb doping — with selenisation (annealing in Se vapour) demonstrating the most significant grain growth, due to the abundant formation of liquid Cu_xSe phases which enhances grain growth by liquid phase sintering.[11] Selective Sb doping at the top layer of CIS creates a high dopant concentration gradient, resulting in a better grain growth due to an increased dopant diffusion. The mechanism of grain growth is similar to the liquid phase sintering of CISSe grains by the formation of low melting point Cu_3SbSe_4 phases along the grain boundaries (Figure 5).

Kesterite-based $Cu_2ZnSnS_xSe_{4-x}$ (CZTSSe) thin film solar cells have made impressive progress for the past few years demonstrating the best efficiency of ~12% which is rather low compared with the performance of $Cu(In,Ga)(S,Se)_2$ (CIGS), because of the intrinsic limitations of CZTSSe, such as secondary phases, anti-site defects, carrier lifetime and so on. Moreover, the efficiency of pure sulfide Cu_2ZnSnS_4 (CZTS, 8.4%) can be improved to 11.3% by the partial substitution of Zn in CZTS films with cadmium forming $Cu_2Zn_{1-x}Cd_xSnS_4$ (CZCTS). This effect of Zn/Cd on cell performance may be attributed to the change in electronic structure of the bulk CZCTS thin film (i.e. phase change from kesterite to stannite), affecting the band

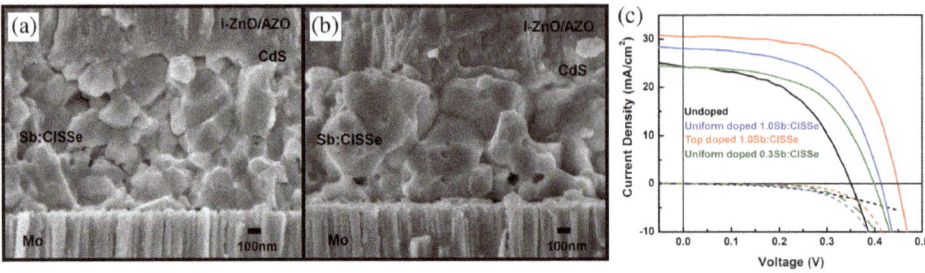

Fig. 5. Cross-sectional FESEM images of CISSe cell with (a) uniform Sb doping, (b) Sb doped at top layer and (c) J-V characteristics of all CISSe cells fabricated (broken lines denote dark currents of the respective configurations).

alignment and the charge separation at the CZCTS/buffer interface. In addition, the ZnS secondary phase can be decreased and the grain sizes can be improved to some degree by the partial replacement of Zn with Cd.

1.5. Organic solar cells

As the organic solar cell is a rapidly increasing technology owing to its flexibility, low weight, low cost of manufacturing, transparency, recyclability and scalability, research on organic solar cells is being conducted by several groups in Singapore, for example at SERIS, at NUS, NTU, and at A*STAR. The Organic Solar Cell (OSC) group at SERIS focuses mainly on bulk heterojunction organic solar cells. In this type of solar cell, the photovoltaic energy conversion follows different mechanisms than in semiconductor solar cells. The main novel characteristics are: donor-acceptor charge carrier generations; selective majority charge carrier transport; and self-assembly of the light conversion structure. The goals of the research in this area are twofold: (i) to develop very cheap flexible solar cells that can be produced by printing technologies and (ii) to conduct use-inspired basic research on solar cells that constitute model systems for a novel class of photovoltaic devices.

For example, SERIS' OSC group has been working on improving the solar cell efficiency through better nanomorphology by employing solution processing to develop a so-called "pseudo bi-layer organic solar cell". These devices are fabricated by the sequential processing of donor, poly(3-hexyl thiophene) (P3HT), and acceptor ([6,6]-phenyl C61-butyric acid methyl ester, PCBM or indene-C60 bisadduct, ICBA) using an orthogonal solvent (the material of the second layer has to be dissolved in an organic solvent that does not dissolve any material of the first layer), followed by subsequent thermal annealing. They have a higher PV efficiency than conventional bulk heterojunction organic solar cells. The best pseudo bi-layer organic solar cell made as yet by SERIS has a 1-Sun efficiency of 5.9%. For more details, the interested reader is referred to the literature.[12] SERIS has also developed nickel oxide (NiO) and molybdenum oxide (MoO_x) hole selective interfacial layers for organic solar cells based on a poly(3-hexylthiophene) (P3HT) donor and a [6,6]-phenyl-C61 butyric acid methyl ester (PCBM) acceptor, prepared by a facile, low-temperature solution process, which enhances the device efficiency and stability in comparison to reference devices containing the conventional (PEDOT:PSS) interfacial layer. Further details can be found in the literature.[13]

Another area of work at SERIS is research on inverted organic solar cells (IOSCs) which display improved device stability compared to the standard organic cells. As the three main aspects of the research at SERIS are: (a) processability, (b) practicality and (c) stability, some of the advances that have been achieved in recent years are shown in Figure 6. For example, one processing step — the UV treatment step prior to the deposition of the hydrophilic PEDOT:PSS layer onto the hydrophobic P3HT:PCBM layer — has been substituted with a new fluorosurfactant, a Capstone FS-31 layer, during the device fabrication process which provides an excellent coating quality while providing decent device efficiency and maintaining the device stability. Furthermore, the light-soaking issue, which significantly hampers the practicality of IOSCs, was resolved by employing fluorinated TiO_x ("F-TiO_x") to replace the conventional TiO_x. Consequently, the light-soaking time was significantly reduced

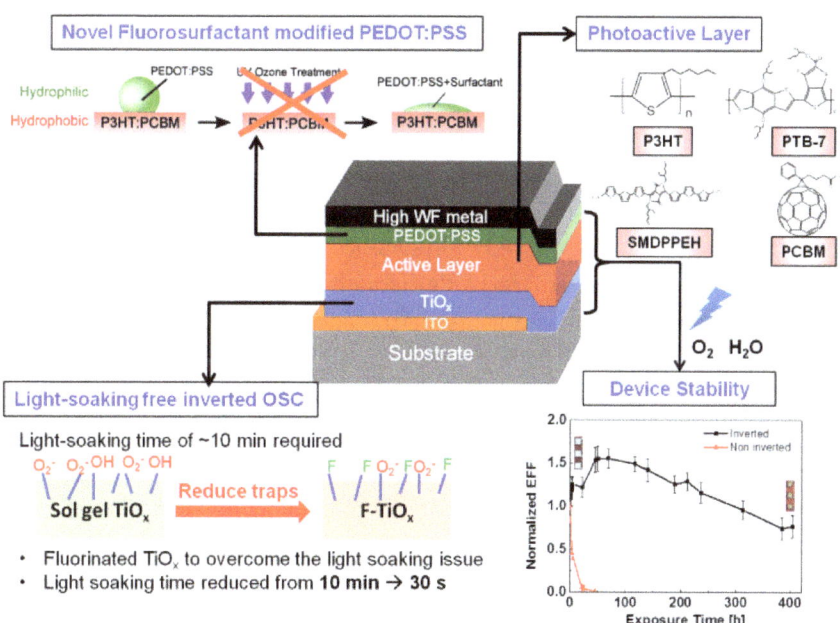

Fig. 6. Research advances at SERIS with inverted organic solar cells, improving the processability, practicality, and device stability and efficiency.

from 10 minutes to 30 seconds by reducing the density of oxygen-induced defects in the material.[14] Lastly, an extensive study on device stability was carried out on the modified IOSCs with a high-efficiency organic donor material (PTB-7). The use of PTB-7 does not alter the processability and the light-soaking time reduction of the final device, suggesting its feasibility to be applied in general IOSC applications.[14]

1.6. Dye solar cells and hybrid solar cells

Further advancements in cost effective and easily processable solar cells are made with the development of Dye Sensitized Solar Cells (DSSCs) where the charge separation in DSSCs is mainly governed by kinetic competition between charge injection, charge recombination, charge transfer and dye and electrolyte regeneration at various interfaces is unlike the Si solar.[15] These solar cells also offer better performance under low light conditions as well as design options for integration into buildings.[15] A best efficiency of 13% under simulated sunlight with AM1.5 spectral distribution has been demonstrated for liquid DSSCs while 10% efficiencies have been achieved in solid state DSSCs with $CsSnI_3$ as an HTM.[16] Though the solid state DSSCs resolved problems associated with the volatility and corrosive nature of the electrolyte, their performance is bottlenecked by poor infiltration of the HTM (hole-transport material), lower optical cross-sections associated with thinner mesoporous films and higher HTM resistance and recombination for thicker mesoporous films.

Research efforts related to DSSCs and hybrid solar cells intensified under the NRF funded CRP project, "Nanonets: New Materials, Devices for Integrated Energy Harnessing & Storage," (NRF-CRP4-2008-3) with Professors from NTU (Subodh Mhaisalkar, Nripan Mathews, Madhavi Srinivasan, Lydia Wong, Lam Yeng Ming), and NUS (Seeram Ramakrishna, Chorng Haur Sow, Wang Qing, and BVR Chowdari). Partnerships were also formed with Professors Michael Grätzel (École Polytechnique Fédérale de Lausanne), Anders Hagfeldt (Ecole Polytechnique de Lausanne), Juan Bisquert (Universitat Jaume I), and Satoshi Uchida (University of Tokyo). The resultant solar cell materials sets and resulting efficiencies are represented in Figure 7. They will be explained briefly in the subsequent paragraphs.

Nanostructured cuprous oxide (Cu_2O, p-type), coupled with ZnO nanorods and hematite (Fe_2O_3, n-type) nanorods have been demonstrated in extremely thin film (ETA) solar cell configurations, an offshoot of DSSC, delivering efficiencies of 0.43% and 0.05% respectively.[17] These materials have been deposited by inexpensive routes such as electrodeposition and hydrothermal methods, and by nanostructuring, the inherent characteristics of these materials has been exploited to the maximum. Additionally light trapping mechanisms — by employing 3D photonic

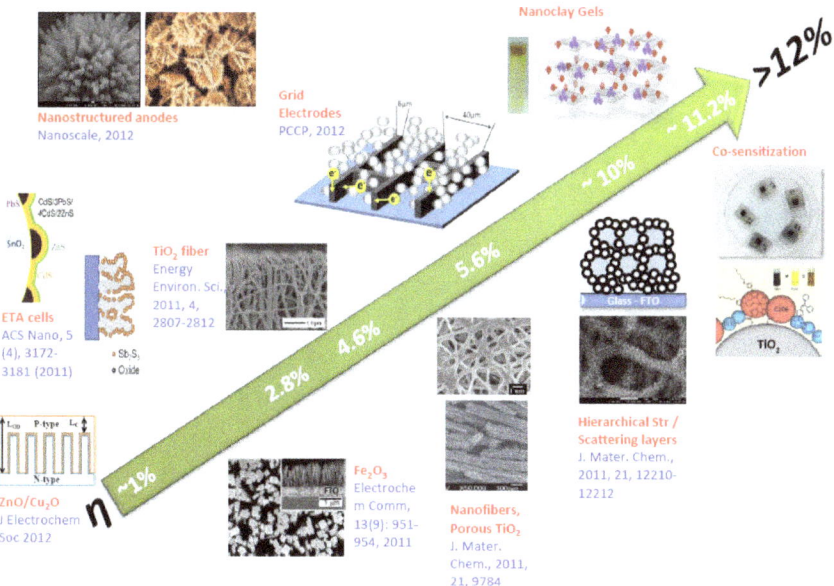

Fig. 7. Schematic showing progress in DSSC performance attained through conscientious strategies and device architectures.

crystals such as TiO$_2$ inverse opals (TIOs, Figure 8a) electrode-fabricated by ALD (2.22% efficiency) and Si nanostructures such as Si nanowires, nanoprisms and nanocones (Figure 8b) — demonstrated how the efficiency of the solar cells can be enhanced. Efficiency increases as the lattice size of the TIO electrode decreases due to the larger surface area for dye loading. Liu et al. was the first to report TiO$_2$ inverse opals (TIO) with a superior infiltration of ca. 96% of the maximum possible infiltration, by conformal filling of 288, 390 and 510 nm opals (Figure 8(1)), giving rise to high quality TIO.[18] The team has recently demonstrated the method to extend this TIO approach for upconversion for enhanced near-infrared light harvesting.[19] Inorganic/organic hybrid solar cells (e.g. Si/PEDOT:PSS, Figure 8(2)) incorporating Si nanostructures have advantages — combining high carrier mobility and excellent light trapping capability of Si with low material cost, low temperature and solution based processability of organic semiconductors yielding solar cells efficiencies in excess of 12%.[20] Nanostructuring is further extended to Zn doped SnO$_2$ nanoflowers and electrospun anatase TiO$_2$ nanofibers which are integrated in liquid DSSCs, yielding power conversion efficiencies of 6.78% and 4.2%.[21] From EIS analysis it was demonstrated that these nanostructures have a better charge collection efficiency, long diffusion lengths, lower resistance to charge transport and higher resistance to recombination of electrons with oxidized species. Scaling down of material costs

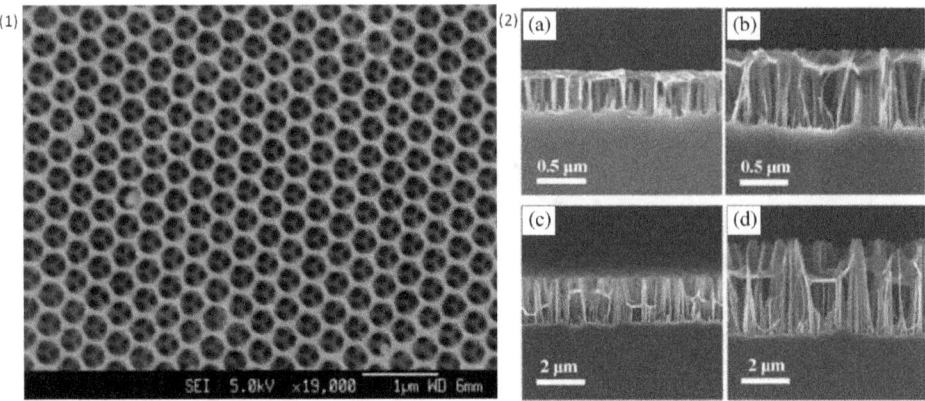

Fig. 8. (1) FESEM of TiO$_2$ Inverse Opals by ALD and (2) cross-sectional SEM images of SiNWs with different lengths of (a) 0.35 mm, (b) 0.9 mm, (c) 2.2 mm and (d) 4.1 mm coated with PEDOT:PSS in Si/PEDOT:PSS hybrid cells.

is also achieved by replacing fluorinated tin oxide (FTO) with 3-dimensional metal grid electrodes (3D-MGEs) as electron collectors in DSSCs — which exhibited 80% transparency and solar conversion efficiencies of 6.2%, which is comparable to the conventional ones but at lower price.[22]

Alternatively, a higher light harvesting efficiency contributing to a higher efficiency has been shown by employing the approach of cosensitization dyes with complementary absorption properties as well as with favourable energy level alignment (for e.g. C106 and D131). Consequently, these devices showed significant enhancement in the efficiency (11.1%), in contrast to devices sensitized with either D131 ($\eta = 5.6\%$) or C106 ($\eta = 9.5\%$).[23] An alternate strategy to enhance the overall solar cell performance is to improve the Voc which can be attained by nitrate-hydrotalcite nanoclay quasi solid state dye-sensitized solar cells, where the nanoclay acts as an additive in the liquid electrolyte to form a gel, and assists in shifting the conduction band of the semiconductor film upwards, giving rise to a higher Voc and efficiencies of >10%.[24] The DSSCs look promising but for commercial applications, efficiencies greater than 15% — attainable by breakthroughs in the areas of efficient materials for absorbers and charge transporting materials — is essential.

1.7. Perovskite solar cells

Perovskites which exhibit the simple AMX$_3$ (cations, A; metal, M; anion, X; Figure 9a) based configuration — and have the propensity for a rich diversity in composition, structure, and properties that include ferroelectricity, superconductivity, semiconducting and catalytic attributes to name a few — are considered the potential materials for photovoltaics due to reports of efficient light

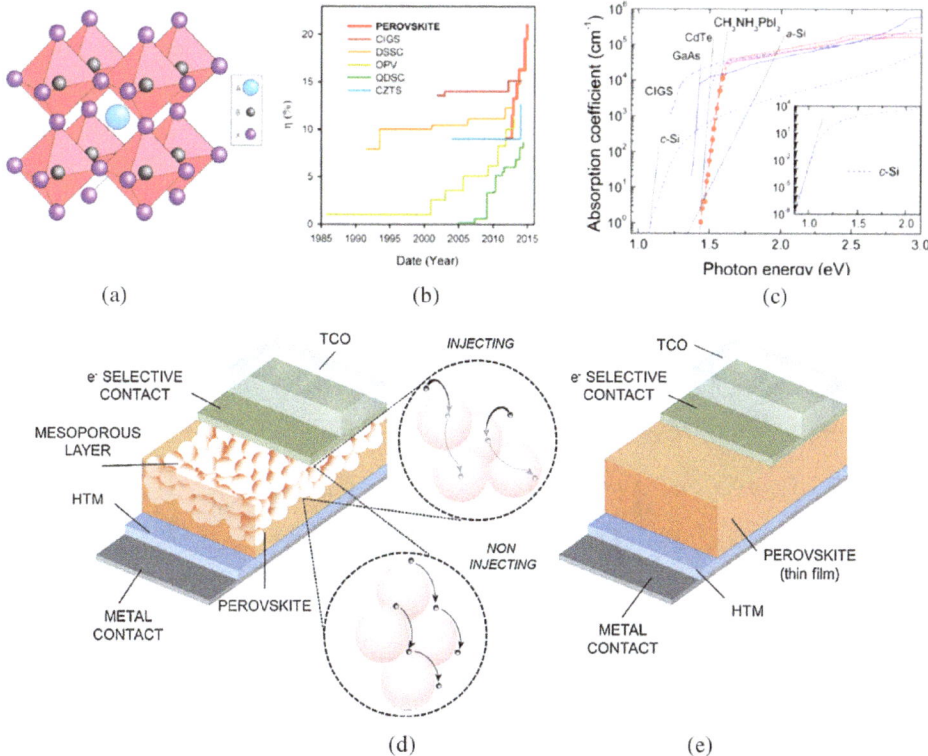

Fig. 9. (a) Perovskite crystal structure, (b) efficiency chart and (c) absorption coefficients for solution processed solar cells. Schematic demonstrating perovskite solar cell architecture — (d) with mesoporous scaffold and (e) thin film configuration.

harvesting and lasing in organometallic halides, especially by methylammonium lead iodide ($CH_3NH_3PbI_3$) and its analogues. Within a span of two years, this organic–inorganic lead halide perovskite had yielded photovoltaic efficiencies of 20.1%, being the highest performing solution-processed solar cell on record and displacing technologies such as dye solar cells and organic photovoltaics that were showing signs of inertia in their development after two decades of intensive research (Figure 9b).

The high efficiencies of the $CH_3NH_3PbI_3$ system were attributed to sharp band gap close to the ideal, high absorbance (Figure 9c), low exciton binding energy with the excited state composed primarily of free carriers, near perfect crystalline film formation with very low defect densities, balanced electron–hole transport, and excellent charge carrier mobilities. The huge interest of $CH_3NH_3PbI_3$ perovskite does not only lie in the high efficiencies but also in the novel configurations made possible by the singular characteristics of the material. It has been shown how the absorber can act as well as an electron transporting layer, as a hole transporting layer, or even as a thin film solar cell (Figure 9d and e). This exciting potential is combined with other advantages. The easy processability, low illumination performance, and versatility position this kind of solar cells as a serious candidate for industrial exploitation. The team at ERI@N had made significant progress in the materials, fundamental property studies, and device architectures of these solar cells and had attained 17.9% efficiency with a Voc of 1.1V, Jsc of 21 mA/cm² and FF of 77.8% (Figure 10a). The progress made since 2012 is summarized below.

The team at ERI@N had demonstrated band gap tuning (Figure 10b),[25] the modification of crystal structure (Figure 11b), morphology evolution, device stabilization, the modulation of film conductivity and thereby efficiency (Figure 10c and d)[26] by tuning the halide composition. Our efforts to be in alignment with green energy led to the development of the first lead-free perovskite based solar cells

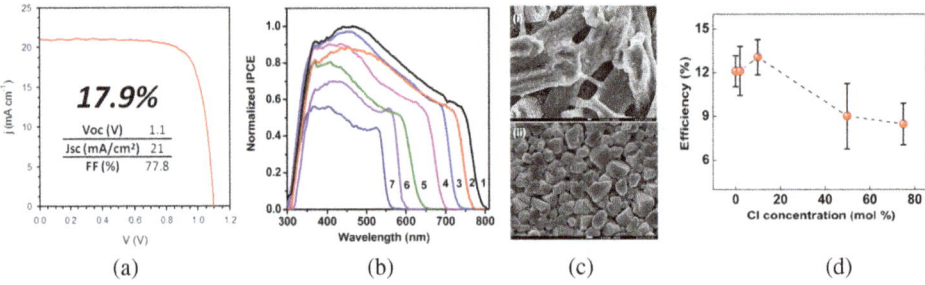

Fig. 10. (a) Champion $CH_3NH_3PbI_3$ solar cell exhibiting 17.9% PCE. (b) Band-gap tuning of $CH_3NH_3PbI_3$ by Br_2 addition. (c) Morphological advancements in $CH_3NH_3PbI_3$ by incorporation of Cl and (d) its impact on PCE for different concentrations of Cl addition.

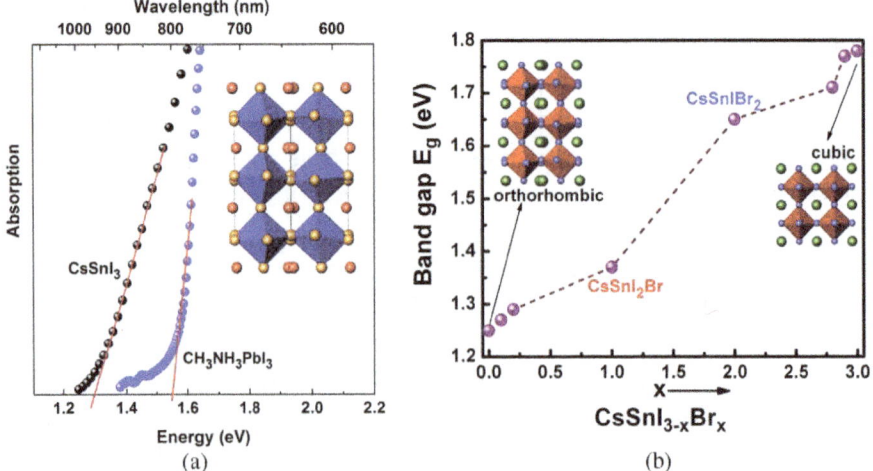

Fig. 11. (a) Crystal structure and band gap of alternative lead-free perovskite — $CsSnI_3$. (b) Band gap tuning of $CsSnI_3$ by Br_2 doping.

displaying very high photocurrent densities of more than 22 mA/cm² with CsSnI$_3$ and HC(NH$_2$)$_2$ SnI$_3$ (FASnI$_3$)[27] owing to their small band gaps being 1.3eV (Figure 11a) and 1.4eV respectively. A very important scientific revelation regarding the addition of SnF$_2$ into Sn based perovskites has been made as it reduces Sn vacancies and renders the Sn perovskite less conductive in nature. This approach opens up new possibilities of exploiting solution processed lead-free perovskites such as CsSnI$_3$ and its organic analogues for use as highly efficient photovoltaics.

The underlying reasons behind these high efficiencies is attributed to their longer electron and hole transport lengths (both over 100nm)[28] (Figure 12d) which was measured by femtosecond transient optical spectroscopy measurements in CH$_3$NH$_3$PbI$_3$ heterojunctions with selective electron and hole extraction layers (Figure 12a). These findings indicate that this class of materials does not suffer from the bottleneck of low collection lengths which handicaps typical low temperature solution processed photovoltaic materials.

This family of organic-inorganic halides CH$_3$NH$_3$PbX$_3$ (where X = Cl, Br, I) were also shown to function as excellent optical gain materials yielding ultra-low thresholds for amplified spontaneous emission (ASE) (Figure 13) stemming from the small bulk defect densities and the insensitivity to the surface traps. Substitution of the halide ions allows for wide wavelength-tunability with the ASE tunable across the entire visible spectrum (390 — 790 nm). These materials with balanced ambipolar charge transport characteristics may lead to realizing electrically-driven lasing with solution processed semiconductors.

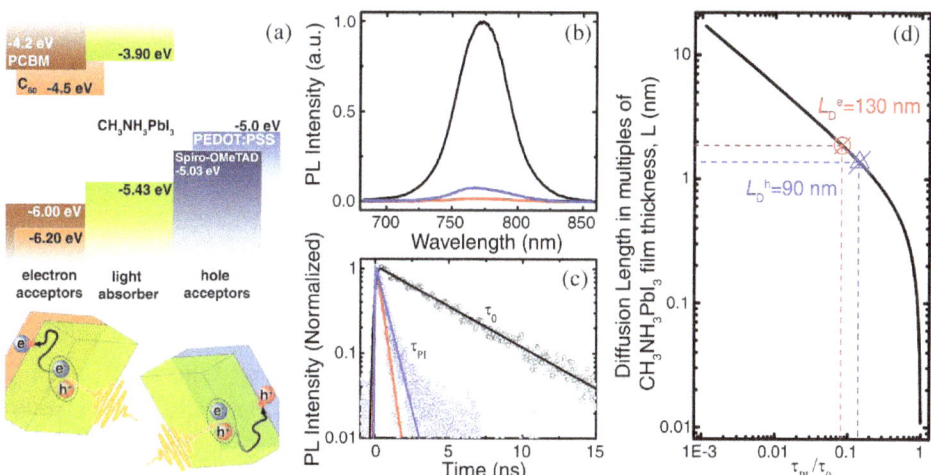

Fig. 12. (a) Schematic of the energy levels of the heterojunctions and depiction of the exciton generation, diffusion and quenching processes in the respective bilayers. (b) Time-integrated PL spectra and (c) Time-resolved PL decay transients for CH$_3$NH$_3$PbI$_3$ alone (black), CH$_3$NH$_3$PbI$_3$ in contact with an electron acceptor (red) and in contact with a hole acceptor (blue). (d) A plot of exciton diffusion length vs PL lifetime quenching ratios.

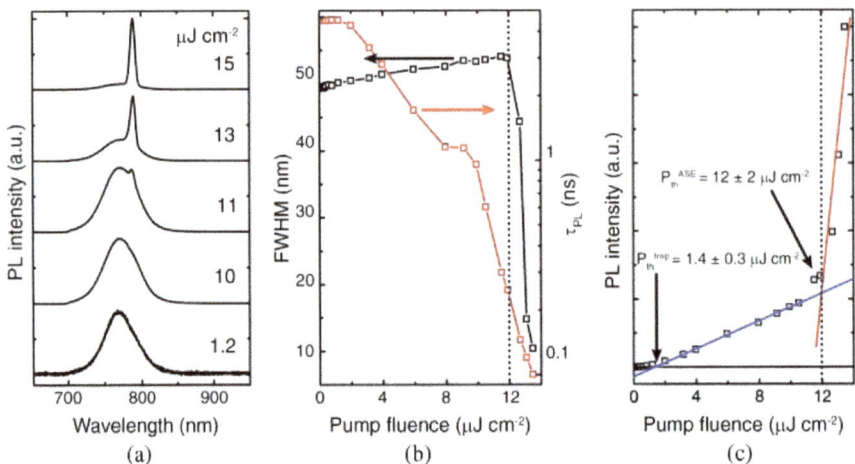

Fig. 13. Coherent light emission from solution processed perovskite film. (a) Steady-state PL emission spectra from a 65 nm thick $CH_3NH_3PbI_3$ film photoexcited using 600 nm, 150 fs and 1 KHz pump pulses with increasing pump fluence (per pulse) — illustrating the transition from SE to ASE. (b) FWHM of the emission peak and average transient PL lifetime (t_{PL}) as a function of the pump fluence. t_{PL} is the time taken for the intensity to decrease to 1/e of its initial value. (c) PL intensity as a function of pump fluence. The arrows indicate the trap state saturation threshold and the ASE threshold. The blue and red lines represent the linear fits to experimental data in the two linear regimes of SE and ASE, respectively.

These desirable inherent characteristics of perovskites have been further investigated in conjunction with nanostructured materials (e.g. sub-micrometre rutile TiO_2 nanorods,[29] electrospun nanofibers[30]) in terms of the infiltration of HTMs and charge collection. Efficiencies of more than 9.4% were demonstrated and it has been inferred that by tailoring the nanostructures' aspect ratios, porosity and coverage, the performance of these solar cells could be fine-tuned by affecting the charge generation efficiency as well as the recombination kinetics as has been inferred by EIS measurements.[29–30]

In the pursuit of realizing flexible and semi-transparent perovskite solar cells for applications in building integrated photovoltaics, the conventional HTM and evaporated metal contact were substituted by inexpensive carbon nanotubes (CNT) (Figure 14b) which provided semi-transparency as well as dual side illuminations as shown in Figure 14a (3.88% power conversion efficiency from the CNT side).[31] Such flexible, semi-transparent perovskite solar cells could also be realized by using ZnO, formed by a room temperature electrodeposition and a chemical bath deposition (CBD) on ITO substrates[32] (Figure 14c). These excellent opto-electronic properties of the perovskites, coupled with effective engineering designs, offer future prospects of the commercialization of efficient and stable solar cells.

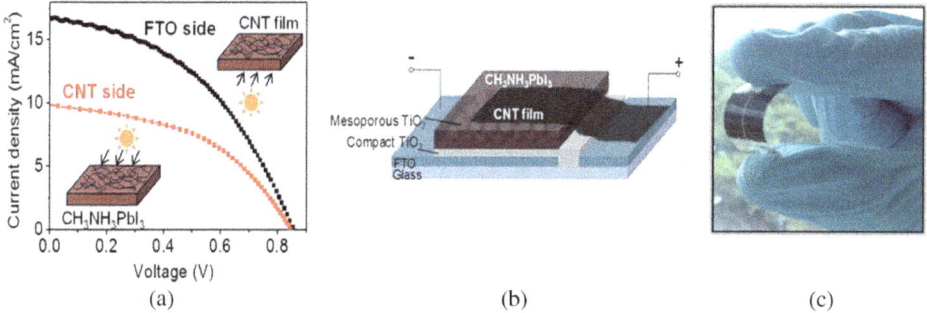

Fig. 14. (a) Light J-V curves of $CH_3NH_3PbI_3$ perovskite/CNTs solar cell with illumination from FTO and from CNT side, under condition of AM1.5 100mW/cm² (b) Schematic of $CH_3NH_3PbI_3$ perovskite solar cell with CNT film electrode. (c) Device on the flexible PET/ITO substrate.

1.8. Future perspective

The future prospects of solar cells look excellent. Crystalline silicon solar cells will remain the workhorse of the global PV industry in the foreseeable future, with continuous improvements of the efficiencies and the costs ($/Wp) of the fabricated PV modules. Using multi- and monocrystalline silicon wafers, PV modules with efficiencies of more than 20% and 24%, respectively, will be commercially available within 5–10 years. Further efficiency improvements towards 30% are expected by 2030, by combining thin-film top cells with silicon wafer bottom cells. The most likely candidates for these thin-film top cells are materials from the III-V, CIGS and perovskite systems. The breakthrough possibility will come from new materials concepts that may include variants of perovskite materials that are yet to be explored for photovoltaics, and alternative device architectures that could include perovskite thin-film tandem cells.

2. Energy Storage Materials Research in Singapore

2.1. Introduction

The energy crisis in recent years, together with the rapid depletion of fossil fuels, increase in world population and environmental pollution, has brought about the need for the sustainable generation and storage of clean energy. Over the past few years, tremendous efforts have been carried out in Singapore to develop clean energy from sources including solar, wind and thermoelectric. Together with these efforts in generating sustainable energy, considerable progress has also been made with the development of advanced energy storage technologies. Among the various energy storage technologies, electrochemical energy storage has been the most promising and has been widely used. In principle, these electrochemical energy storage devices store energy in the format of electricity, which is made possible through electrochemical processes by the charge and discharge of electrons and ions, such as Li^+, Na^+, K^+, H^+ and OH^-. A typical electrochemical energy storage device is composed of the following three indispensable components: an anode, a cathode, and an electrolyte. During the discharge process, ions will de-sorb/extract from one electrode and transfer to the other one, at the same time the electron thus generated flows through an external circuit; in the reverse, during the charge process the device will be under an external potential across the electrodes, thus the reverse electrochemical reactions (ions adsorption/insertion) take place at the electrodes. After the charge process, the electrochemical device will become polarized with the energy stored. The charge storage mechanisms of both batteries and supercapacitors are different, for example batteries undergo the Faradaic (electron transfer between electrode electrolyte) reaction whereas the supercapacitors' process is a non-Faradaic reaction i.e. physical adsorption/de-sorption. This fundamental difference in the mechanism results in a high energy density of batteries and a high power density of supercapacitors. The hybridization of both supercapacitor and battery is also an interesting approach to achieve high energy and power density. This part briefly overviews the research activity of Singapore based institutes towards aforesaid electrochemical energy storage devices like batteries, supercapacitors and the hybridization of these two. The background of all the three electrochemical energy storage devices is briefly described in their respective sections.

2.2. Supercapacitor

As one of the most widely used electrochemical energy storage devices, supercapacitors have drawn considerable interest in recent years in the international research community, including Singapore, where the output in research has increased dramatically. Supercapacitors are also termed electrochemical capacitors or ultracapacitors, which in principle can provide a high power density, long lifecycle, safety and device reliability. In addition, they can effectively bridge the energy/power density gap between the common battery (with high energy density) and traditional dielectric capacitor (with high power density).

On the basis of the charge storage mechanisms, supercapacitors are commonly divided into the following two categories: electric double-layer capacitors (EDLCs) and pseudocapacitors.[33] In EDLCs, charges are stored through the physical adsorption of ions, and no chemical reaction is involved during the charge/discharge process, thus the capacitance of the EDLCs is largely determined by the electrostatic accumulation of charges at the electrode/electrolyte interface. In EDLCs, the surface charges can be generated in several ways: surface dissociation, ion adsorption from the electrolyte and surface defects. The charge-discharge process involved in a typical EDLC is as follows: during the charge process with an external potential, electrons move from the negative electrode to the positive electrode. At the same time, positive cations (such as Li^+, Na^+, K^+, H^+) move towards the negative electrode, and negative anions (such as OH^-, Cl^-) move towards the positive electrode within the electrolyte. During the discharge, these processes take place in reverse. In the process, no electrochemical reaction has occurred, thus the energy is stored in the double-layer interface.[34] As the electric energy is stored on the surface of electrode materials, the capacitance is therefore proportional to the electrode surface area and the specific capacitance, C ($F\ g^{-1}$) is given by the following equation:[35]

$$C = \varepsilon_0 \varepsilon_r \frac{S}{D}$$

where ε_0 is the electric constant (8.854 × 10^{12} $F\ m^{-1}$), ε_r is the relative dielectric constant of the interface, S is the specific surface area of the electrode ($m^2\ g^{-1}$), and D (m) is the separation of the electrode plates. The electrode materials for EDLCs are largely carbonaceous materials, such as active carbon (AC), powdered graphite, mesoporous carbon, carbon nanotubes (CNTs) and more recently graphene.[36] Considerable potential has been demonstrated in achieving electrode materials with large and tunable pore volume and high surface area.[37]

On the other hand, pseudocapacitors (also named as Faradaic supercapacitors) rely on the reversible Faradaic reactions of certain transition metal oxides and hydroxides (redox reactions) to store energy.[38] The main difference of pseudocapacitors from EDLCs is that chemical reactions take place on the electrode surface during the charge/discharge process. Since the electrochemical processes occur on both the electrode surface and subsurface region of the electrode, a pseudocapacitor usually exhibits a much larger capacitance value and therefore higher energy density than that of an EDLC.[39] However, it can greatly suffer from the limited active surface area, relatively slow redox reactions at the electrode as well as low electrical conductivity, thus the overall power density and cycling ability may not be comparable with that of an EDLC. From a materials point of view, electrode materials for pseudocapacitors can be divided into conducting polymers (such as polypyrrole and polyaniline) and metal oxides/hydroxides (such as RuO_2, MnO_2, Co_3O_4, and $Ni(OH)_2$). In order to realize high performance pseudocapacitors, the electrode materials should essentially exhibit a large surface area and high electrical conductivity.

2.2.1 Carbon-based materials

As typical EDLC electrode materials, carbon-based materials — with a tunable pore volume and pore size distribution and high surface area, high electrical conductivity, and desired electrochemical stability — have been developed.[35] They are also abundant in nature with hence a relatively lower cost.

So far, the most widely used carbon materials include activated carbon, mesoporous carbon, carbon aerogels, carbon nanofibers, carbon nanotubes (CNTs) and graphene. The electrochemical behaviour of activated carbons is largely related to their porous structure and surface functionalities.[43] The desirable carbon structure should have a properly controlled pore size distribution with most pore sizes matching the size of the electrolyte.

Wang's research group at the National University of Singapore reported the fabrication of meso-/nano-porous carbon that exhibit highly ordered mesoporous structures with an electrochemical capacitance of 290 F g^{-1}, as well as activated carbon using KOH displaying 316 F g^{-1} with excellent cycling stability.[44] They were developed by a hydrothermal method, as shown in Figure 15a[39] and environmentally friendly carbohydrate precursors and Pluronic F127 (for soft template) were used as the carbon source. By controlling the hydrothermal temperature and duration, pore configuration, pore size distribution and specific surface area of the porous carbon could be varied.

High specific surface area materials such as graphene (2620 m² g^{-1})[45] and graphene based materials which exhibit excellent electrical conductivity, mechanical strength, flexibility and stability are deemed to be excellent electrodes for supercapacitors.[46] So rGO films which are paper-like, lightweight, porous, continuously crosslinked and electrically conductive have been reported by Chen's group (shown in Figure 15b)[40] to exhibit a specific capacitance of about 110 F g^{-1} for a two-electrode cell. Several research efforts from Nanyang Technological University include a 3D graphene network developed by chemical vapour deposition (CVD), for which ethanol was used as the carbon source and commercial Nickel foam as a template[41] (shown in Figure 15c), and hydrothermally grown Co_3O_4 nanowires on the 3D graphene (Figure 15d)[42] resulting in 3D graphene/Co_3O_4, which can be used as a flexible free-standing electrode in supercapacitors. For the former case, graphene was firstly coated on Ni foam and then by removing the template, a high-quality 3D graphene network can be obtained. The graphene substrate can be further coated with metal oxide (e.g. NiO) to achieve enhanced energy density and a specific capacitance of 816 F g^{-1} with excellent electrochemical stability. While the 3D graphene/Co_3O_4 showed a very high specific capacitance of ~1100 F g^{-1} at a high current density of 10 A g^{-1}, together with excellent cycling stability.

Fig. 15. Supercapacitor electrode materials: (a) porous carbon; Adapted with permission from ref.[39], Copyright© 2013 Royal Society of Chemistry. (b) Graphene paper; Adapted with permission from ref.[40], Copyright© 2012 Wiley-VCH. (c) 3D graphene; Adapted with permission from ref.[41], Copyright© 2011 Wiley-VCH. (d) 3D graphene/Co_3O_4 composite; Adapted with permission from ref.[42], Copyright© 2012 American Chemical Society. Insert in (b) shows the digital image of a flexible supercapacitor made from graphene paper.

2.2.2. Metal oxides and hydroxides

Transition metal oxides/hydroxides in principle can deliver a high energy density when used as supercapacitor electrode materials, since the oxidation states are available for redox reactions. The metal oxides based electrodes which have been studied extensively include RuO_2, IrO_2, MnO_2, NiO, Co_3O_4 and $Ni(OH)_2$, $Co(OH)_2$.

Research groups in NTU and NUS have demonstrated the synthesis and application of various metal oxides such as sub-10 nm diameter-MnO_2 nanowires (Figure 16a) with a specific capacitance of 279 F g^{-1} and desired cycling stability,[47] the Ni foam-mesoporous the Co_3O_4 nanosheet (Figure 16b)[20a] with ultrahigh specific capacitances in the range of 2735–1471 F g^{-1} with excellent cycling stability, and hollow high surface area nanocarbon spheres.[48] Upon coating of the carbon sphere with a thin layer of MnO_2, the carbon-MnO_2 nanohybrid spheres demonstrated

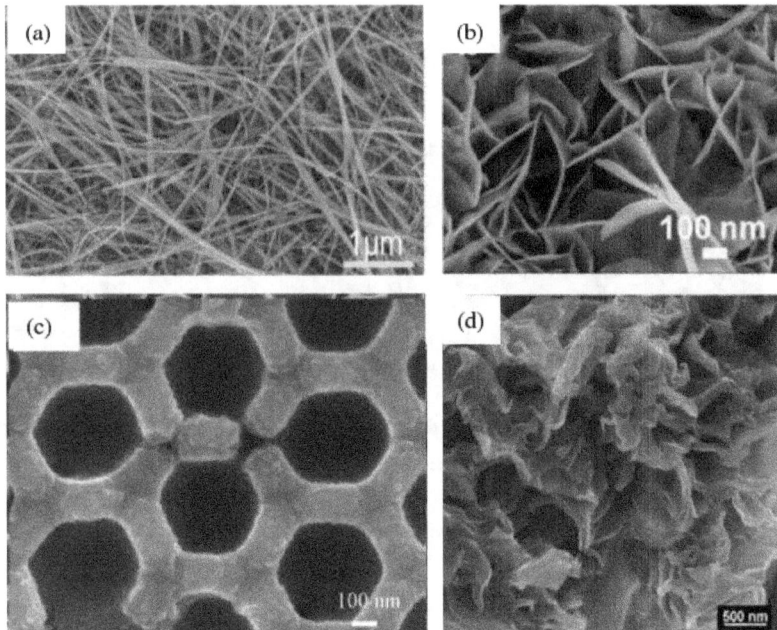

Fig. 16. Supercapacitor electrode materials: (a) MnO_2 nanowires; Adapted with permission from ref.[47], Copyright© 2011 Royal Society of Chemistry. (b) Co_3O_4 nanosheet; Adapted with permission from ref.[20a], Copyright© 2012 Royal Society of Chemistry. (c) PANI/carbon; Adapted with permission from ref.[52], Copyright© 2009 American Chemical Society. (d) PANI/graphene; Adapted with permission from ref. [51], Copyright© 2012 Royal Society of Chemistry.

a promising application for use as supercapacitors. These materials could be synthesized by the simple routes of either hydrothermal or electrodeposition.

2.2.3. Conducting polymers

Conducting polymers can store energy by redox reactions, thus their energy densities are also higher than those of carbonaceous materials.[49] Compared with metal oxides and hydroxide, which are relatively expensive, conducting polymers are much cheaper. Generally, a conducting polymer supercapacitor can give rise to a much higher energy density (10 Wh kg^{-1}) than that of a carbonaceous based one. However, the swelling and shrinking of conducting polymers during the intercalating/de-intercalating process can result in mechanical degradation resulting in a poor cycling stability in comparison to carbonaceous materials and metal oxides.[49] As typical electrode materials, the widely used conducting polymer materials are polyaniline (PANI), polypyrrole, polythiophene and derivatives of polythiophene.[50]

Research groups from the Department of Chemical and Biomolecular Engineering as well as from the Department of Chemistry at NUS have demonstrated composite electrode material comprising of PANI with macroporous carbon (Figure 16c) or graphene (Figure 16d), yielding very high specific capacitances of 1490 F g^{-1} and 526 F g^{-1} at a current density of 0.2 A g^{-1} respectively, with excellent rate performance and cycle ability. A proper combination of the unique carbon matrix and the high pseudocapacitance of the PANI layer, and a homogeneous dispersion of graphene sheets and polymer matrix achieved — due to the desired dispensability of surfactant-stabilized graphene in aqueous phase — were the origins for the excellent performance displayed by this composite electrode.[51]

2.2.4. Hierarchical electrode materials

Hierarchical electrode materials which have high surface area, are nanostructured, have active two or more materials, exhibit advanced supercapacitor electrode capability with high performance as they provide more reaction sites with the electrolyte, thereby shortening the electron pathway.[53]

Fan's group at Nanyang Technological University reported a cost-effective strategy to produce hierarchical hybrid nanostructure arrays for supercapacitor application, which is shown in Figure 17a.[54] The integrated electrode made of a Co_3O_4@MnO_2 core/shell nanowire array exhibited excellent electrochemical performance, which is ascribed to the synergetic contribution from the porous Co_3O_4 nanowire core and the ultrathin MnO_2 shell. This pioneer work opened up the possibility of constructing high-performance pseudocapacitive materials with different functional materials. The same group also reported a novel wire-in-tube structure for a supercapacitor.[55] With

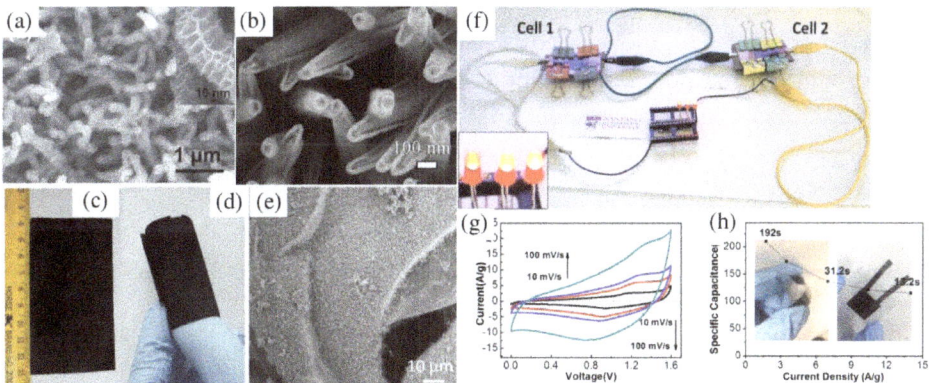

Fig. 17. Supercapacitor electrode materials: (a) Co_3O_4@MnO_2; Adapted with permission from ref.[54], Copyright© 2011 Wiley-VCH. (b) CoO-TiO_2; Adapted with permission from ref.[55], Copyright© 2012 Royal Society of Chemistry. (c-e) 3D graphene-CNT film, (f) image of the GF/CNT/MnO_2//GF/CNT/Ppy supercapacitor; Adapted with permission from ref.[57], Copyright© 2014 Royal Society of Chemistry. (g-h) Electrochemical test results of the GF/CNT/Fe_2O_3//GF/CoMoO$_4$ full cell; Adapted with permission from ref.[58], Copyright© 2015 American Chemical Society.

the help of atomic layer deposition, they demonstrated a general concept of the "gapped core–shell nanostructure" for electrochemical energy storage application, as shown in Figure 17b. The "wire in tube" and "wall in box" structures of $CoO–TiO_2$ and $NiO-TiO_2$ have been fabricated and tested as supercapacitor electrode materials. The nanogapped hollow core–shell electrodes give rise to a much higher capacitance than the solid core–shell nanorods as the former supplies a larger reaction area facilitating the electrolyte contact with the active material while the outer tube layer preserves the structural integrity. While the above two examples discuss metal oxides, yet also promising is the idea of combining metal oxides with carbonaceous materials and/or conducing polymers to achieve synergetic effects, thus obtaining high performance supercapacitor electrodes.[56]

Ke and co-workers in the Materials Science and Engineering Dept at NUS reported a 3D hierarchical $SnO_2@Ni(OH)_2$ core-shell nanowire array, which was directly grown on a flexible carbon cloth.[58] Due to the synergetic efforts from both the metal oxides and hydroxide, the structure can achieve a high specific capacitance of 1553 F g^{-1}, and a high capacitance of 934 F g^{-1} can be maintained even when the discharge current increased 20 times, demonstrating its excellent rate capability.

Shen's research group at Nanyang Technological University reported a novel graphene foam/carbon nanotube film, which could be used as a conductive and flexible substrate for the deposition of MnO_2 and polypyrrole to form flexible asymmetric $GF/CNTs/MnO_2$ and GF/CNTs/Ppy hybrid films (as shown in Figure 17c-f) which can produce an output voltage of 1.6 V and a high energy/power density (22.8 Wh kg^{-1} at 860 W kg^{-1} and 2.7 kW kg^{-1} at 6.2 Wh kg^{-1}) with excellent rate property and cycling ability. Using the GF/CNT film as a substrate, researchers at NUS, Nanyang Poly, and NTU have demonstrated the deposition of ultrafine iron oxide nanoparticles on the surface of the carbon substrate (forming one system), as shown in Figure 17g-h,[58] exhibiting outstanding electrochemical performances. When combined with a graphene/$CoMoO_4$ cathode, the supercapacitor device could deliver a high energy of ~74.7 Wh kg^{-1} at a power of 1.4 kW kg^{-1}, together with excellent cycling ability. A similar work has been reported for Co_3O_4 coated CNTs with carbon cloth,[53d] which also showed good supercapacitor performance, revealing the merits of the hierarchical electrode material for use in supercapacitors.

2.3. Lithium ion batteries (LIB)

Lithium-ion batteries (LIB) still remain popular and are excellent portable electrochemical energy storage devices in this era. The LIB power packs are intensively pursued for electric (EV), hybrid electric (HEV) and plug-in electric vehicles (P-HEV) applications. Since the commercialization of — LIB in the "rocking-chair" configuration with graphite as the anode and $LiCoO_2$ as the cathode — for consumer applications by Sony in 1990 which dominated the entire electronic appliances market for over two decades, they still remain popular at present. Chowdari's research group from NUS was one of the early groups to commence LIB materials research in Singapore. After that several research groups in Singapore including Madhavi (NTU), Yan (NTU), Lou (NTU), Palani Balaya (NUS) and Seeram (NUS) have contributed significantly to the advancement of LIB.

The development of high energy density LIB completely depends on the utilization of high voltage cathodes and carbonaceous anodes (e.g. graphite) which essentially act as a buffer medium during the cycling process. Among several candidates such as $LiNi_{0.5}Mn_{1.5}O_4$,[59] $LiCoPO_4$,[60] $Li_3V_2(PO_4)_3$,[61] V_2O_5,[62] $LiCoO_2$,[63] $LiMn_2O_4$, $LiFePO_4$[64] proposed for LIB applications, spinel and its derivatives were found appealing in terms of their high working potential, eco-friendliness, long-term cycleability and high power capability. In the case of spinels, 3D pathways for the Li-ion migration is certainly beneficial compared to the other reported cathodes. Similarly graphitic electrodes are dominant as anodes in most of the commercial LIB, though however they pose drawbacks like electrolyte decomposition in the first discharge (which leads to the formation of solid electrolyte interface, SEI) and Li-platting at a high current operation. Therefore, intense research work was carried out in the field of anodes. Generally, the anodes used in LIB are broadly classified into three main categories viz. intercalation, conversion (displacement), and alloy types based on the reaction mechanism with Li. Several transition metal oxides have been proposed as Li-insertion hosts, for example, $LiCrTiO_4$,[65] TiP_2O_7,[66] $LiTi_2(PO_4)_3$,[67] $TiNb_2O_7$,[68] Nb_2O_5,[69] $Li_4Ti_5O_{12}$,[70] and TiO_2.[71] Similarly, displacement materials (such as Fe_2O_3, Fe_3O_4, NiO etc.), and alloying materials (such as Si, Sn, Ge etc.) with various morphological aspects[72] were extensively investigated by the group of Madhavi, Yan and Lou of NTU. Nevertheless, most of the electrode studies were limited to half-cell performance only,[71j] which was insufficient to understand the real time application of these materials. Full cell LIB evaluation is more practical and most of these studies have been reported by Madhavi's group at NTU (Figure 18). Electrospun TiO_2 nanofiber, when coupled with $LiMn_2O_4$ cathode, rendered ~81% of initial reversible capacity after 100 cycles with an operating potential of ~2.2 V.[71h] On the other hand, full-cell, $LiMn_2O_4/TiO_2$ with PVdF-HFP nanofiber membrane delivered outstanding cycleability of 700 cycles with capacity retention of ~90% (Figure 18).[71e] These results clearly indicate the influence of 1D morphology and its favourable electrochemical activity even in harsh conditions.[71h] Furthermore, the energy density/operating potential is substantially improved by replacing the $LiMn_2O_4$ cathode by high voltage $LiNi_{0.5}Mn_{1.5}O_4$ using the same electrospinning approach to form an all 1D nanostructured LIB.[73] This $LiNi_{0.5}Mn_{1.5}O_4/TiO_2$ cell operates at ~2.8 V with a reversible capacity of ~102 mAh g^{-1}. Furthermore, the full-cell retained ~86% of initial reversible capacity after 400 cycles, which is much better than the

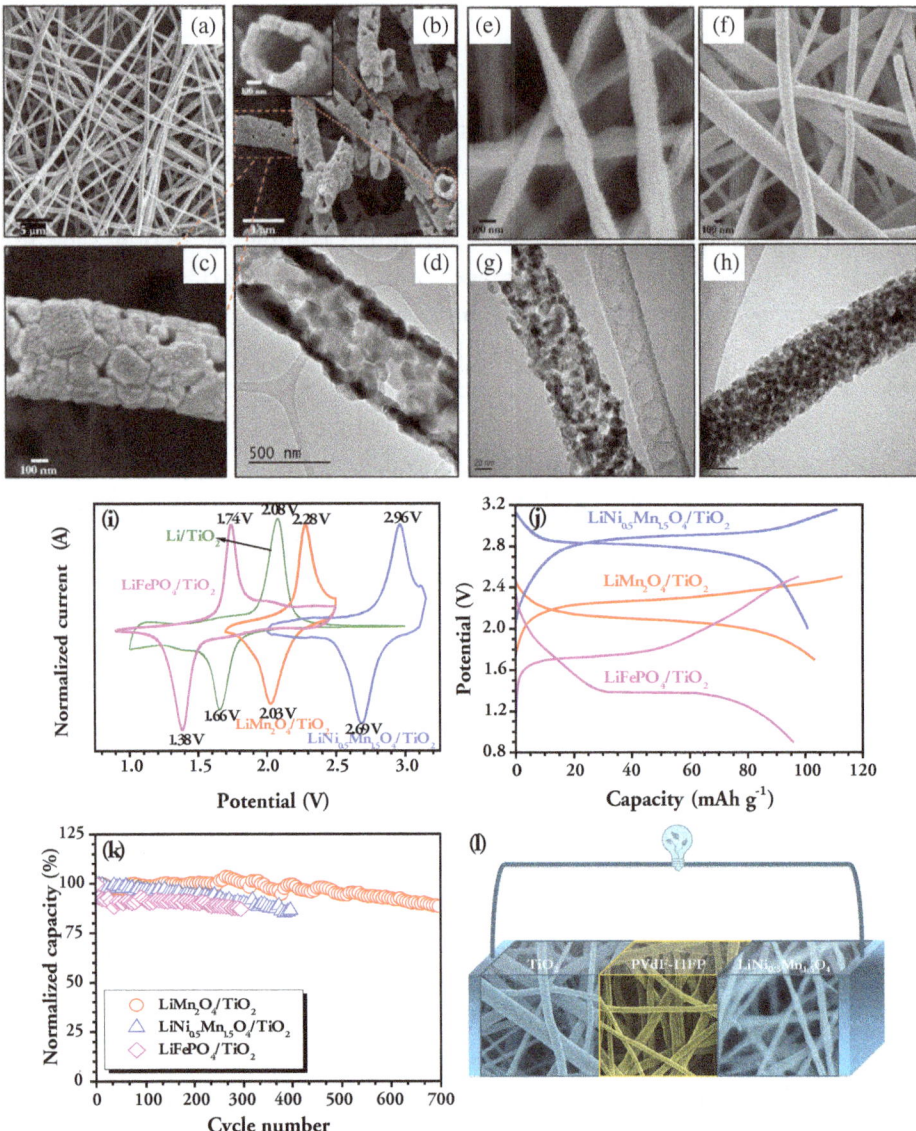

Fig. 18. (a) FE-SEM image of as-spun $LiMn_2O_4$ nanofibers (green nanofibers); (b) FE-SEM image of electrospun $LiMn_2O_4$ hollow nanofibers sintered at 800 °C for 5 h. Inset: magnified view of nanofibers indicating hollow structure; (c) Magnified view of the single hollow nanofibers; (d) TEM image of sintered electrospun $LiMn_2O_4$ hollow nanofibers; FE-SEM images of calcined nanofibers of (e) $LiNi_{0.5}Mn_{1.5}O_4$, (f) anatase TiO_2; TEM images of calcined nanofibers (g) $LiNi_{0.5}Mn_{1.5}O_4$ and (h) anatase TiO_2; (i) Typical CV signatures of anatase TiO_2 anodes in full-cell assemblies with eco-friendly cathodes: $LiMn_2O_4$ (red line), $LiNi_{0.5}Mn_{1.5}O_4$ (blue line), and $LiFePO_4$ (pink line) at a slow 0.1 mV s^{-1} scan rate. The performance of anatase TiO_2 in a half-cell assembly (Li/TiO_2) is also given (green dashed line) for comparison; (j) Typical galvanostatic charge-discharge curves of $LiMn_2O_4/TiO_2$ (red line, current density: 150 mA g^{-1}), $LiNi_{0.5}Mn_{1.5}O_4/TiO_2$ (blue line, current density: 15 mA g^{-1}), and $LiFePO_4/TiO_2$ (pink line, current density: 100 mA g^{-1}) cells; (k) Plot of the normalized reversible capacity of the aforementioned cells relative to the number of cycles; (l) Schematic representation of a typical LIB composed of all 1D-nanostructured components.[68, 75]

previously reported work by Brutti et al.[74] This result clearly reveals and parallels the importance of going the "nano-concept" route to realize such electro-active materials in unique 1D architecture made by electrospinning. Furthermore, studies have been carried out to realize the "nano-concept" route by altering the high capacity $TiNb_2O_7$ anodes in an all 1D $LiMn_2O_4/TiNb_2O_7$ configuration resulting in an operating potential of ~2.4 V which is ~0.2 V higher than the former configuration and hence with a slightly higher net energy density.

In addition to materials research in LIB, a state of the art prototyping facility had been built to construct a prototype of cylindrical (18650) and laminated LIB prototype under the flagship of Energy research institute at NTU (ERI@N). Several successful overseas university and industry collaborations had been set up by ERI@N NTU in Singapore under the energy storage area. Some of the prominent ones are TUM CREATE, BMW, JM, Sud-Chemie, SGL Carbon, Bosch, Gildemeister.

2.4. Li — ion hybrid electrochemical capacitors (Li-HEC)

Though LIB offers high energy density and good cycleability, it seriously suffers from being unable to deliver a high power to drive the EV and HEV. Hence, a high power electrochemical energy storage device is desperately required to drive the aforesaid vehicles. In the recent past, research focus has been directed towards the development of high performance lithium ion (Li-ion) hybrid electrochemical supercapacitors (Li-HEC) because of their higher power density than the rechargeable battery (e.g. LIB) and higher energy density than supercapacitors (e.g., electric double-layer capacitors, EDLC). Thus, the Li-HEC is expected to bridge the gap between the LIB and EDLCs and become the ultimate power source for the aforesaid applications in the near future (Figure 19a).

These Li-HEC are a new type of electrochemical energy storage device which exhibits the features of both supercapacitor and LIB and is anticipated to demonstrate higher energy and power densities. The research group of Madhavi, NTU has been working on the development of the Li-HECs in Singapore with focus on the new Li-insertion type and double layer forming high surface area carbonaceous electrodes especially from the bio-mass derived products, such as coconut shell, waste paper, etc.

The Li-HEC utilizes both Faradaic and non-Faradaic processes to store charge to achieve higher energy density than the EDLC, and higher power density than LIB without sacrificing cycling stability. Therefore, high surface area carbonaceous materials were chosen as the capacitor type electrode materials (for non-Faradaic process) for Li-HEC, by keeping long term cycleability and cost effectiveness in mind. On the other hand, a high performance Li-insertion type electrode material was necessary to deliver a high energy density during prolonged cycling.

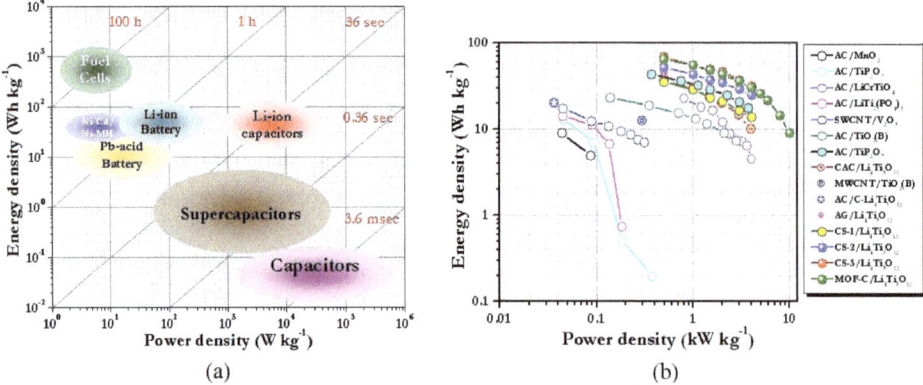

Fig. 19. (a) Ragone plot shows the demand for higher power/energy electrochemical energy storage devices relative to present day technology.[70a] (b) Ragone plot indicates the performance of various insertion type anode materials tested for Li-HEC applications in non-aqueous medium. AC: activated carbon, MWCNT: multi-walled carbon nanotube, SWCNT: single walled carbon nanotube, CS-2: coconut shell granules are treated with $ZnCl_2$ in the presence of CO_2, CS-1: coconut shell granules in hydrothermal treatment and pyrolysed in the presence of CO_2 in succession, CS-3: coconut shell granules in hydrothermal reaction and pyrolysed in the presence of CO_2 with addition of $ZnCl_2$, MOF-C: metal organic framework derived carbon.

Unfortunately, only a few materials have been explored as insertion type electrode materials for Li-HEC applications. Hence, the search for high performance Li-insertion type electrode materials is imperative and so the materials tested for Li-HEC applications in non-aqueous medium had been reviewed. On the other hand, the utilization of aqueous electrolyte for Li-HEC application had been discontinued, owing to the poor compatibility of the medium with insertion electrodes and importantly the limited operating potential (~1.23 V is the decomposition potential of the water).

For the fabrication of non-aqueous Li-HEC, insertions type anode materials such as TiP_2O_7, $LiTi_2(PO_4)_3$, $LiNi_{0.5}Mn_{1.5}O_4$, $TiNb_2O_7$, $Li_2Ti_3O_7$, V_2O_5, $LiCrTiO_4$, α-MnO_2, TiO_2-B and $Li_4Ti_5O_{12}$ (exploited by the group of Madhavi) and cathode materials such as $LiFePO_4$, $LiCoPO_4$, $LiMn_2O_4$, $LiNi_{0.5}Mn_{1.5}O_4$, $LiNi_{1/3}Mn_{1/3}Fe_{1/3}O_2$, $LiNi_{1/3}Mn_{1/3}Co_{1/3}O_2$, Li_2MoO_3, Li_2MnSiO_4, Li_2FeSiO_4 and Li_2CoPO_4F have been proposed to construct a high performance Li-HEC with carbonaceous materials as the counter electrode. In contrast to non-Faradaic electrodes, insertion type electrodes can accommodate Li-ions at various potentials with little variation in the voltage as a function of state of charge. Proper utilization of the insertion type compounds by optimizing the mass loading can result in a dramatic increase of the average output voltage and subsequent enhancement in the energy density of the Li-HEC as well. Insertion type compounds are very crucial to construct a high performance Li-HEC because they not only provide the enhancement in energy and power densities but also exhibit exceptional robustness and safety features, so rational design is critical.

Apparent from the Ragone plot (Figure 19), it is clear that there are three materials, TiO_2 anatase and bronze phases $LiCrTiO_4$ and $Li_4Ti_4O_{12}$ found appealing and dominant compared to the rest of the insertion type anodes investigated for Li-HEC applications. Amongst spinel phase materials, $Li_4Ti_4O_{12}$ exhibits favourable properties for application in a Li-HEC. However, there are still a few issues especially pertaining to the power density to be addressed. To address these, conductive coatings or pre-lithiated graphite anode with a high surface area AC anode were proposed. However such coatings reduce volumetric capacitance and increase production cost while in the latter case, pre-lithiation results in the formation of SEI over the surface which consumes a huge amount of Li during the first discharge resulting in an increase in internal resistance of the system. These effects hinder the achievement of a desired power density. Other insertion type anode spinel phase $LiCrTiO_4$, which exhibits slightly lower theoretical capacity than $Li_4Ti_5O_{12}$ for reversible Li-insertion was not explored much due to the toxicity of the Cr. Although the higher energy density was observed for pristine (~30 Wh kg^{-1}) and the anatase TiO_2-rGO composite (~42 Wh kg^{-1}), the ICL (irreversible capacity loss), higher operating potential and poor long-term cycleability were issues to deeply consider for employing them in practical devices compared to spinel anodes.[76] In order to overcome these issues, another polymorph of TiO_2, the bronze phase was introduced, which exhibited lower operating potential and a negligible ICL rendering it as a prospective anode material for Li-ion batteries and Li-HEC applications. Performance of TiO_2-B as an anode in Li-HEC was found comparable to that of both spinel phase $LiCrTiO_4$ and $Li_4Ti_5O_{12}$ anodes with the AC counter electrode. Moreover, cycleability is another important issue for the TiO_2-B polymorph. This clearly showed that the spinel phase $Li_4Ti_5O_{12}$ decorated with carbonaceous materials was the best choice to be used as an insertion type electrode used in Li-HEC along with either a high surface area AC or graphene. Similar to the insertion type anodes, there were several Li-insertion type cathode active materials explored for Li-HEC applications. The energy densities of insertion type cathode materials were better than anodes in the Li-HEC configuration, which was attributed to the higher redox potential of the transition metal oxide used. Long term cycleability is very critical for Li-HEC, but achieving such cycleability is questionable for the case of cathodes, since most of them reported here are based on "Mn" which is soluble in the conventional carbonate based electrolytes in the Jahn-Teller region. This clearly showed that for an advancement in the insertion type electrode, focus on insertion type anodes, rather than cathodes, is crucial. Therefore, high capacity with low operating potential anodes are highly anticipated to fulfil the necessary demand to satisfy the requirements to drive zero emission transportation applications.

In Li-HEC, most of the research works are devoted to the development of an insertion type electrode, unfortunately not much work has been carried out on cathodic materials. Nevertheless, the group of Madhavi exploited a few carbonaceous counter electrodes derived from various cost effective and eco-friendly sources like coconut shell, waste paper, polymers, metal organic framework (MOF) etc. Generally, AC has been utilized as the unanimous choice for the counter electrode in a Li-HEC assembly. Although it has unique properties like high specific surface area, electronic conductivity, thermal stability, excellent chemical stability with wide range of pH values and cost effectiveness, performance characteristics of such an electrode in aqueous and non-aqueous medium were entirely different.[77] The AC obtained from the chemical activation in the hydrothermal carbonization and the subsequent physical activation process (CS-ZHTP) delivered a maximum energy density of ~69 Wh kg^{-1}. It was generally believed that chemical activation processes at high temperature suppresses electrical conductivity of the carbonaceous materials while hydrothermal carbonization process not only overcame these issues but also generated higher percentage of mesoporosity during subsequent pyrolysis (~60%).[70b] Furthermore without any chemical activation, hydrothermal and physical activation processes yielded (CS-HTP) AC delivering superior power capability and a similar energy density in comparison to commercial AC. Hence, hydrothermal activation plays a vital role in attaining a high power density. Interestingly CS-ZP prepared with chemical and physical activation processes alone translated a higher energy density (~52 Wh kg^{-1}) than the commercial AC without any hydrothermal treatment. This clearly suggests that the activation process is significant for the appropriate tuning of energy densities. Irrespective of the supercapacitor configuration (symmetric, asymmetric and hybrid) and electrolyte medium (aqueous or non-aqueous) are very crucial. Also, graphene based systems performed better than conventional high surface area porous carbons and the AC electrode. However, the variation in batch to batch graphene synthesis posed challenges to its reproducibility, reliability and scalability.[78] As such we believe that AC derived from bio-mass precursors is the best choice for real time applications of Li-HEC.

2.5 Future Perspective

For the past several years, excellent progress has been achieved in Singapore, in the area of energy storage, as evidenced by the impact of research and papers published. Singapore is a key player in the international community. Moving ahead, there is great potential for further research progress in energy storage. To realize the goal, there are several challenges in order to meet the ever-increasing demand for both stationary and portable devices, which require high energy and power densities, as well as device stability and reliability. Since the performance of batteries and supercapacitors (e.g., capacity, power density, rate capability and cycling ability) is greatly determined by the intrinsic properties of electrode materials, proper design and rational synthesis of advanced electrode materials is the key challenge for developing the next generation of high performance energy storage device. The research efforts in Singapore will pave a way for commercialized energy storage devices with high energy/power density, long-time cycling stability, low-cost, flexibility and safety.

Significant progress in the development of insertion type anodes has been realized in NTU and NUS in all the three categories, and development has matured enough as well. Recently, spinel $Li_4Ti_5O_{12}$ anode based cells have been commercialized, shortly after the graphite based negative electrodes. The insertion type, bronze phase and $TiNb_2O_7$ are much favourable for commercialization. Nevertheless, there have been several alloy type anodes which make potential candidates and are expected to replace the conventional graphitic anodes, but the unusual volume variation and associated cyclic stability renders hinder the practical cells. Therefore, much research works are anticipated to overcome the issues described above and to be realized for the fabrication of higher energy density Li-ion power packs with lighter weight and higher power capability, eventually powering the EV and HEV in the near future. Similarly for the case of Li-HEC, much more intense work is needed for the development of carbonaceous electrodes to realize higher power capability without compromising the energy density.

Acknowledgements

The Solar Energy Research Institute of Singapore (SERIS) is sponsored by the National University of Singapore (NUS) and Singapore's National Research Foundation (NRF) through the Singapore Economic Development Board (EDB). The Energy Research Institute @ NTU (ERI@N) is supported by NTU, NRF, EDB and other government agencies in Singapore. John Wang acknowledges the support of Dr Guan Cao in drafting the section of Energy Storage Materials, without whom, it would be impossible to put this together.

References

1. Kanti Basu, P., Z. Hameiri, D. Sarangi, J. Cunnusamy, E. Carmona, M.B. Boreland, 18.7% Efficient inline-diffused screen-printed silicon wafer solar cells with deep homogeneous emitter etch-back. *Solar Energy Materials and Solar Cells*, 2013. 117, 412–420.

2. Basu, P.K., M.B. Boreland, V. Shanmugam and D. Sarangi, PCT patent application PCT/SG2013/000183 2013, 2013.
3. Wong, J. In *Proc. 39th IEEE Photovoltaic Specialists Conference*, Tampa Bay, IEEE, New York: Tampa Bay, 2013.
4. Kuwano, K. S.T., M. Onishi, H. Nishikawa, S. Nakano and T. Imai, *Jpn. J. Appl. Phys.* 1980. **20**(213).
5. Lechner, P. and H. Schade, Photovoltaic thin-film technology based on hydrogenated amorphous silicon. *Progress in Photovoltaics: Research and Applications*, 2002. **10**(2): pp. 85–97.
6. Staebler, D.L. and C.R. Wronski, Reversible conductivity changes in discharge produced amorphous Si. *Applied Physics Letters*, 1977. **31**(4): pp. 292–294.
7. Matsuyama, T., N. Terada, T. Baba, T. Sawada, S. Tsuge, K. Wakisaka and S. Tsuda, High-quality polycrystalline silicon thin film prepared by a solid phase crystallization method. *Journal of Non-Crystalline Solids*,1996. 198–200, Part 2, 940–944.
8. Keevers, M.J., T. L. Y., U. Schubert and M.A. Green, Proc. 22nd European Photovoltaic Solar Energy Conf., Milan, WIP, Munich: Milan, 2007; p. 1783.
9. Kumar, A., P. I. Widenborg, G. K. Dalapati, G. S. Subramanian and A.G. Aberle, Impact of deposition parameters on the material quality of SPC poly-Si thin films using high-rate PECVD of a-Si:H. *EPJ Photovolt.* 2015. 6, 65303.
10. Meier, J. S.D., R. Flückiger, D. Fischer, H. Keppner and A. Shah, Proc. 1stWorld Conf. on Photovoltaic Energy Conversion, Hawaii, IEEE, New York: Hawaii, 1994; p. 409.
11. Xiang, Y., X. Zhang and S. Zhang, Insight into the mechanism of Sb promoted Cu(In,Ga)Se2 formation. *Journal of Solid State Chemistry*, 2013. 204, 278–282.
12. (a) Heinemann, M.D., K. Ananthanarayanan, L.N.S.A. Thummalakunta, C.H. Yong and J. Luther, *GREEN The International Journal of Sustainable Energy Conversion and Storage*, 2011. 1, 291–298; (b) Thummalakunta, L.N.S.A., C.H. Yong, K. Ananthanarayanan and J. Luther, *Organic Electronics*, 2012. 13, 2008–2016.
13. Wong, K.H., K. Ananthanarayanan, M.D. Heinemann, J. Luther and P. Balaya, Enhanced photocurrent and stability of organic solar cells using solution-based NiO interfacial layer. *Solar Energy*, 2012. **86**(11): pp. 3190–3195.
14. Lim, F.J., A. Krishnamoorthy and G.W. Ho, Device Stability and Light-Soaking Characteristics of High-Efficiency Benzodithiophene–Thienothiophene Copolymer-Based Inverted Organic Solar Cells with F-TiO$_x$ Electron-Transport Layer. *ACS Applied Materials & Interfaces*, 2015. **7**(22): pp. 12119–12127.
15. O'Regan, B. and M. Gratzel, A low-cost, high-efficiency solar cell based on dye-sensitized colloidal TiO$_2$ films. *Nature*, 1991. **353**(6346): pp. 737–740.
16. (a) Chung, I., B. Lee, J. He, R.P.H. Chang and M.G. Kanatzidis, All-solid-state dye-sensitized solar cells with high efficiency. *Nature* 2012, 485 (7399), 486–489; (b) Mathew, S., A. Yella, P. Gao, R. Humphry-Baker, F.E. CurchodBasile, N. Ashari-Astani, I. Tavernelli, U. Rothlisberger, K. NazeeruddinMd and M. Grätzel, Dye-sensitized solar cells with 13% efficiency achieved through the molecular engineering of porphyrin sensitizers. *Nat Chem*, 2014. **6**(3): pp. 242–247.

17. (a) Yantara, N., N. Mathews, K.B. Jinesh, H.K. Mulmudi and S.G. Mhaisalkar, Modulating the optical and electrical properties of all metal oxide solar cells through nanostructuring and ultrathin interfacial layers. *Electrochimica Acta* 2012, 85, 486–491; (b) Mulmudi, H.K., N. Mathews, X.C. Dou, L.F. Xi, S.S. Pramana, Y.M. Lam and S.G. Mhaisalkar, Controlled growth of hematite (α-Fe_2O_3) nanorod array on fluorine doped tin oxide: Synthesis and photoelectrochemical properties. *Electrochemistry Communications*, 2011. **13**(9): pp. 951–954.
18. Liu, L., S.K. Karuturi, L.T. Su and A.I.Y. Tok, TiO_2 inverse-opal electrode fabricated by atomic layer deposition for dye-sensitized solar cell applications. *Energy & Environmental Science*, 2011. **4**(1): pp. 209–215.
19. Su, L.T., S.K. Karuturi, J. Luo, L. Liu, X. Liu, J. Guo, T.C. Sum, R. Deng, H.J. Fan, X. Liu and A.I.Y. Tok, Photon Upconversion in Hetero-nanostructured Photoanodes for Enhanced Near-Infrared Light Harvesting. *Advanced Materials*, 2013. **25**(11): pp. 1603–1607.
20. (a) Yuan, C., L. Yang, L. Hou, L. Shen, X. Zhang and X.W. Lou, Growth of ultrathin mesoporous Co_3O_4 nanosheet arrays on Ni foam for high-performance electrochemical capacitors. *Energy & Environmental Science* 2012, 5, 7883–7887; (b) Wang, J., H. Wang, A.B. Prakoso, A.S. Togonal, L. Hong, C. Jiang and Rusli, High efficiency silicon nanowire/organic hybrid solar cells with two-step surface treatment. *Nanoscale*, 2015. **7**(10): pp. 4559–4565.
21. (a) Sabba, D., N. Mathews, J. Chua, S.S. Pramana, H.K. Mulmudi, Q. Wang and S.G. Mhaisalkar, High-surface-area, interconnected, nanofibrillar TiO_2 structures as photoanodes in dye-sensitized solar cells. *Scripta Materialia*, 2013. **68**(7): pp. 487–490; (b) Dou, X., D. Sabba, N. Mathews, L.H. Wong, Y.M. Lam and S. Mhaisalkar, Hydrothermal Synthesis of High Electron Mobility Zn-doped SnO2 Nanoflowers as Photoanode Material for Efficient Dye-Sensitized Solar Cells. *Chemistry of Materials*, 2011. **23**(17): pp. 3938–3945.
22. Chua, J., N. Mathews, J.R. Jennings, G. Yang, Q. Wang and S.G. Mhaisalkar, Patterned 3-dimensional metal grid electrodes as alternative electron collectors in dye-sensitized solar cells. *Physical Chemistry Chemical Physics*, 2011. **13**(43): pp. 19314–19317.
23. Nguyen, L.H., H.K. Mulmudi, D. Sabba, S.A. Kulkarni, S.K. Batabyal, K. Nonomura, M. Gratzel and S.G. Mhaisalkar, A selective co-sensitization approach to increase photon conversion efficiency and electron lifetime in dye-sensitized solar cells. *Physical Chemistry Chemical Physics*, 2012. **14**(47): pp. 16182–16186.
24. Wang, X., S.A. Kulkarni, B.I. Ito, S.K. Batabyal, K. Nonomura, C.C. Wong, M. Grätzel, S.G. Mhaisalkar and S. Uchida, Nanoclay Gelation Approach toward Improved Dye-Sensitized Solar Cell Efficiencies: An Investigation of Charge Transport and Shift in the TiO_2 Conduction Band. *ACS Applied Materials & Interfaces*, 2013. **5**(2): pp. 444–450.
25. Kulkarni, S.A., T. Baikie, P.P. Boix, N. Yantara, N. Mathews and S. Mhaisalkar, Band-gap tuning of lead halide perovskites using a sequential deposition process. *Journal of Materials Chemistry A*, 2014. **2**(24): pp. 9221–9225.

26. (a) Dharani, S. H.A. Dewi, R.R. Prabhakar, T. Baikie, C. Shi, D. Yonghua, N. Mathews, P.P. Boix and S.G. Mhaisalkar, Incorporation of Cl into sequentially deposited lead halide perovskite films for highly efficient mesoporous solar cells. *Nanoscale*, 2014. **6**(22): pp. 13854–13860; (b) Yantara, N., F. Yanan, C. Shi, H.A. Dewi, P.P. Boix, S.G. Mhaisalkar and N. Mathews, Unravelling the Effects of Cl Addition in Single Step $CH_3NH_3PbI_3$ Perovskite Solar Cells. *Chemistry of Materials*, 2015. **27**(7): pp. 2309–2314.
27. (a) Koh, T.M., T. Krishnamoorthy, N. Yantara, C. Shi, W.L. Leong, P.P. Boix, A.C. Grimsdale, S.G. Mhaisalkar and N. Mathews, Formamidinium tin-based perovskite with low Eg for photovoltaic applications. *Journal of Materials Chemistry A*, 2015. **3**(29): pp. 14996–15000; (b) Kumar, M.H., S. Dharani, W.L. Leong, P.P. Boix, R.R. Prabhakar, T. Baikie, C. Shi, H. Ding, R. Ramesh, M. Asta, M. Graetzel, S.G. Mhaisalkar and N. Mathews, Lead-Free Halide Perovskite Solar Cells with High Photocurrents Realized Through Vacancy Modulation. *Advanced Materials*, 2014. **26**(41): pp. 7122–7127; (c) Sabba, D. H.K. Mulmudi, R.R. Prabhakar, T. Krishnamoorthy, T. Baikie, P.P. Boix, S. Mhaisalkar and N. Mathews, Impact of Anionic Br⁻ Substitution on Open Circuit Voltage in Lead Free Perovskite ($CsSnI_{3-x}Br_x$) Solar Cells. *The Journal of Physical Chemistry C*, 2015. **119**(4): pp. 1763–1767.
28. Xing, G., N. Mathews, S. Sun, S.S. Lim, Y.M. Lam, M. Grätzel, S. Mhaisalkar and T.C. Sum, Long Range Balanced Electron and Hole Transport Lengths in Organic-Inorganic $CH_3NH_3PbI_3$. *Science*, 2013. **342**(6156): pp. 344–347.
29. Kim, H.S., J.W. Lee, N. Yantara, P.P. Boix, S.A. Kulkarni, S. Mhaisalkar, M. Grätzel and N.G. Park, High Efficiency Solid-State Sensitized Solar Cell-Based on Submicrometer Rutile TiO_2 Nanorod and $CH_3NH_3PbI_3$ Perovskite Sensitizer. *Nano Letters*, 2013. **13**(6): pp. 2412–2417.
30. Dharani, S., H.K. Mulmudi, N. Yantara, Thu P.T. Trang, N.G. Park, M. Graetzel, S. Mhaisalkar, N. Mathews and P.P. Boix, High efficiency electrospun TiO_2 nanofiber based hybrid organic-inorganic perovskite solar cell. *Nanoscale*, 2014. **6**(3): pp. 1675–1679.
31. Li, Z., S.A. Kulkarni, P.P. Boix, E. Shi, A. Cao, K. Fu, S.K. Batabyal, J. Zhang, Q. Xiong, L.H. Wong, N. Mathews and S.G. Mhaisalkar, Laminated Carbon Nanotube Networks for Metal Electrode-Free Efficient Perovskite Solar Cells. *ACS Nano*, 2014. **8**(7): pp. 6797–6804.
32. Kumar, M.H., N. Yantara, S. Dharani, M. Graetzel, S. Mhaisalkar, P.P. Boix and N. Mathews, Flexible, low-temperature, solution processed ZnO-based perovskite solid state solar cells. *Chemical Communications*, 2013. **49**(94): pp. 11089–11091.
33. Zhao, X., B.M. Sanchez, P.J. Dobson and P.S. Grant, The role of nanomaterials in redox-based supercapacitors for next generation energy storage devices. *Nanoscale*, 2011. **3**(3): pp. 839–855.
34. (a) Zhang, H., G. Cao, Y. Yang and Z. Gu, Comparison Between Electrochemical Properties of Aligned Carbon Nanotube Array and Entangled Carbon Nanotube Electrodes. *Journal of The Electrochemical Society*, 2008. **155**(2): pp. K19–K22; (b) Pech, D., M. Brunet, H. Durou, P. Huang, V. Mochalin, Y. Gogotsi, P.-L. Taberna

and P. Simon, Ultrahigh-power micrometre-sized supercapacitors based on onion-like carbon. *Nat Nano*, 2010. **5**(9): pp. 651–654.
35. Wang, G., L. Zhang and J. Zhang, A review of electrode materials for electrochemical supercapacitors. *Chem. Soc. Rev.* 2012. **41**(2): pp. 797–828.
36. (a) Sun, H., X. You, J. Deng, X. Chen, Z. Yang, J. Ren and H. Peng, Novel Graphene/Carbon Nanotube Composite Fibers for Efficient Wire-Shaped Miniature Energy Devices. *Advanced Materials* 2014. **26**(18): pp. 2868–2873; (b) Zhang, L. L., X. Zhao, M.D. Stoller, Y. Zhu, H. Ji, S. Murali, Y. Wu, S. Perales, B. Clevenger and R.S. Ruoff, Highly Conductive and Porous Activated Reduced Graphene Oxide Films for High-Power Supercapacitors. *Nano Letters*, 2012. **12**(4): pp. 1806–1812.
37. (a) Lota, G., K. Fic and E. Frackowiak, Carbon nanotubes and their composites in electrochemical applications. *Energy & Environmental Science*, 2011. **4**(5): pp. 1592–1605; (b) Xu, C., B. Xu, Y. Gu, Z. Xiong, J. Sun and G. Zhao, Graphene-based electrodes for electrochemical energy storage. *Energy & Environmental Science*, 2013. 6, 1388–1414.
38. Augustyn, V., P. Simon and B. Dunn, Pseudocapacitive oxide materials for high-rate electrochemical energy storage. *Energy & Environmental Science*, 2014. 7, 1597–1614.
39. Liu, H., Y. Zhang, Q. Ke, K.H. Ho, Y. Hu and J. Wang, Tuning the porous texture and specific surface area of nanoporous carbons for supercapacitor electrodes by adjusting the hydrothermal synthesis temperature. *Journal of Materials Chemistry A*, 2013. **1**(41): pp. 12962–12970.
40. Niu, Z., J. Chen, H.H. Hng, J. Ma and X. Chen, _A Leavening Strategy to Prepare Reduced Graphene Oxide Foams. *Advanced Materials*, 2012. 4144–4150.
41. Cao, X., Y. Shi, W. Shi, G. Lu, X. Huang, Q. Yan, Q. Zhang and H. Zhang, Preparation of Novel 3D Graphene Networks for Supercapacitor Applications. *Small*, 2011. **7**(22): pp. 3163–3168.
42. Dong, X.-C., H. Xu, X.-W. Wang, Y.-X. Huang, M.B. Chan-Park, H. Zhang, L.-H. Wang, W. Huang and P. Chen, 3D Graphene–Cobalt Oxide Electrode for High-Performance Supercapacitor and Enzymeless Glucose Detection. *ACS Nano*, 2012. **6**(4): pp. 3206–3213.
43. Zhai, Y., Y. Dou, D. Zhao, P.F. Fulvio, R.T. Mayes and S. Dai, _Carbon Materials for Chemical Capacitive Energy Storage. *Advanced Materials*, 2011. **23**(42): pp. 4828–4850.
44. Mao, L., Y. Zhang, Y. Hu, K.H. Ho, Q. Ke, H. Liu, Z. Hu, D. Zhao and J. Wang, Activation of sucrose-derived carbon spheres for high-performance supercapacitor electrodes. *RSC Advances*, 2015. **5**(12): pp. 9307–9313.
45. Cao, X., Z. Yin and H. Zhang, Three-Dimensional Graphene Materials: Preparation, Structures and Application in Supercapacitors. *Energy & Environmental Science*, 2014. 7, 1850–1865.
46. Zhang, L.L., R. Zhou and X.S. Zhao, Graphene-based materials as supercapacitor electrodes. *Journal of Materials Chemistry*, 2010. **20**(29): pp. 5983–5992.
47. Jiang, H., T. Zhao, J. Ma, C. Yan and C. Li, Ultrafine manganese dioxide nanowire network for high-performance supercapacitors. *Chemical Communications*, 2011. **47**(4): pp. 1264–1266.

48. Yang, Z.-C., C.-H. Tang, H. Gong, X. Li and J. Wang, Hollow spheres of nanocarbon and their manganese dioxide hybrids derived from soft template for supercapacitor application. *Journal of Power Sources*, 2013. **240**(0): pp. 713–720.
49. Snook, G.A., P. Kao and A.S. Best, Conducting-polymer-based supercapacitor devices and electrodes. *Journal of Power Sources*, 2011. **196**(1): pp. 1–12.
50. Wang, K., H. Wu, Y. Meng and Z. Wei, Conducting Polymer Nanowire Arrays for High Performance Supercapacitors. *Small*, 2014. **10**(1): pp. 14–31.
51. Mao, L., K. Zhang, On H.S. Chan and J. Wu, Surfactant-stabilized graphene/polyaniline nanofiber composites for high performance supercapacitor electrode. *Journal of Materials Chemistry*, 2012. **22**(1): pp. 80–85.
52. Zhang, L.L., S. Li, J. Zhang, P. Guo, J. Zheng and X.S. Zhao, Enhancement of Electrochemical Performance of Macroporous Carbon by Surface Coating of Polyaniline. *Chemistry of Materials* 2009, **22**(3): pp. 1195–1202.
53. (a) Guan, C., X. Li, Z. Wang, X. Cao, C. Soci, H. Zhang and H.J. Fan, Nanoporous Walls on Macroporous Foam: Rational Design of Electrodes to Push Areal Pseudocapacitance. *Advanced Materials*, 2012. 24(30): pp. 4186–4190; (b) Guan, C., X. Li, H. Yu, L. Mao, L.H. Wong, Q. Yan and J. Wang, A novel hollowed CoO-in-CoSnO$_3$ nanostructure with enhanced lithium storage capabilities. *Nanoscale*, 2014. **6**(22): pp. 13824–13830; (c) Reddy, A.L.M., S.R. Gowda, M.M. Shaijumon and P.M. Ajayan, Hybrid Nanostructures for Energy Storage Applications. *Advanced Materials*, 2012. **24**(37): pp. 5045–5064; (d) Guan, C., Y. Wang, M. Zacharias, J. Wang and H.J. Fan, Atomic-layer-deposition alumina induced carbon on porous Ni$_x$Co$_{1-x}$O nanonets for enhanced pseudocapacitive and Li-ion storage performance. *Nanotechnology*, 2015. **26**(1), 014001; (e) Jiang, J., Y. Li, J. Liu, X. Huang, C. Yuan and X.W. Lou, Recent Advances in Metal Oxide-based Electrode Architecture Design for Electrochemical Energy Storage. *Advanced Materials*, 2012. 24, 5166–5180; (f) Tang, Z., C.-h. Tang and H. Gong, A High Energy Density Asymmetric Supercapacitor from Nano-architectured Ni(OH)$_2$/Carbon Nanotube Electrodes. *Advanced Functional Materials*, 2012. **22**(6): pp. 1272–1278; (g) Wang, X., X. Li, X. Sun, F. Li, Q. Liu, Q. Wang and D. He, Nanostructured NiO electrode for high rate Li-ion batteries. *Journal of Materials Chemistry*, 2011. **21**(11): pp. 3571–3573.
54. Liu, J., J. Jiang, C. Cheng, H. Li, J. Zhang, H. Gong and H.J. Fan, Co$_3$O$_4$ Nanowire MnO$_2$ Ultrathin Nanosheet Core/Shell Arrays: A New Class of High-Performance Pseudocapacitive Materials. *Advanced Materials*, 2011. **23**(18): pp. 2076–2081.
55. Guan, C., X. Xia, N. Meng, Z. Zeng, X. Cao, C. Soci, H. Zhang and H.J. Fan, Hollow core-shell nanostructure supercapacitor electrodes: gap matters. *Energy & Environmental Science*, 2012. 5, 9085–9090.
56. Cheng, C. and H.J. Fan, Branched nanowires: Synthesis and energy applications. *Nano Today*, 2012. **7**(4): pp. 327–343.
57. Liu, j., L.L. Zhang, H. Wu, J. Lin, Z. Shen, D.X.-W. Lou, High-performance Flexible Asymmetric Supercapacitors Based on A New Graphene Foam/Carbon Nanotubes Hybrid Film. *Energy & Environmental Science*, 2014. 7, 3709–3719.

58. Ke, Q., C. Guan, M. Zheng, Y. Hu, K.-h. Ho and J. Wang, 3D hierarchical SnO_2@$Ni(OH)_2$ core-shell nanowire arrays on carbon cloth for energy storage application. *Journal of Materials Chemistry A*, 2015. **3**(18): pp. 9538–9542.
59. (a) Kim, M.C., K.-W. Nam, E. Hu, X.-Q. Yang, H. Kim, K. Kang, V. Aravindan, W.-S. Kim, Y.-S. Lee, Sol–Gel Synthesis of Aliovalent Vanadium Doped $LiNi_{0.5}Mn_{1.5}O_4$ Cathodes with Excellent Performance at High Temperatures. *ChemSusChem*, 2014. **7**(3): pp. 829–834; (b) Arun, N., A. Jain, V. Aravindan, S. Jayaraman, W. Chui Ling, M.P. Srinivasan and S. Madhavi, Nanostructured spinel $LiNi_{0.5}Mn_{1.5}O_4$ as new insertion anode for advanced Li-ion capacitors with high power capability. *Nano Energy*, 2015. **12**(0): pp. 69–75.
60. Yang, S.M.G., V. Aravindan, W.I. Cho, D.R. Chang, H.S. Kim and Y.S. Lee, Realizing the Performance of $LiCoPO_4$ Cathodes by Fe Substitution with Off-Stoichiometry. *Journal of The Electrochemical Society*, 2012. **159**(7): A1013–A1018.
61. (a) Rui, X., Q. Yan, M. Skyllas-Kazacos and T.M. Lim, $Li_3V_2(PO_4)_3$ cathode materials for lithium-ion batteries: A review. *Journal of Power Sources*, 2014. 258(0): pp19–38; (b) Son, J.N., S.H. Kim, M.C. Kim, G.J. Kim, V. Aravindan, Y.G. Lee and Y.S. Lee, Superior charge-transfer kinetics of NASICON-type $Li_3V_2(PO_4)_3$ cathodes by multivalent Al^{3+} and Cl^- substitutions. *Electrochimica Acta*, 2013. 97, 210–215; (c) Cho, A.R., J.N. Son, V. Aravindan, H. Kim, K.S. Kang, W.S. Yoon, W.S. Kim and Y.S. Lee, Carbon supported, Al doped-$Li_3V_2(PO_4)_3$ as a high rate cathode material for lithium-ion batteries. *Journal of Materials Chemistry*, 2012. **22**(14): pp. 6556–6560.
62. (a) Cheah, Y. L., V. Aravindan and S. Madhavi, Improved Elevated Temperature Performance of Al-Intercalated V_2O_5 Electrospun Nanofibers for Lithium-Ion Batteries. *ACS Applied Materials & Interfaces*, 2012. **4**(6): pp. 3270–3277; (b) Cheah, Y. L., V. Aravindan, Madhavi, S., Synthesis and Enhanced Lithium Storage Properties of Electrospun V_2O_5 Nanofibers in Full-Cell Assembly with a Spinel $Li_4Ti_5O_{12}$ Anode. *ACS Applied Materials & Interfaces*, 2013. **5**(8): pp. 3475–3480; (c) Aravindan, V., Y.L. Cheah, W.F. Mak, G. Wee, B.V.R. Chowdari, S. Madhavi, Fabrication of High Energy-Density Hybrid Supercapacitors Using Electrospun V_2O_5 Nanofibers with a Self-Supported Carbon Nanotube Network. *ChemPlusChem*, 2012. **77**(7): pp. 570–575; (d) Cheah, Y.L., N. Gupta, S.S. Pramana, V. Aravindan, G. Wee and M. Srinivasan, Morphology, structure and electrochemical properties of single phase electrospun vanadium pentoxide nanofibers for lithium ion batteries. *Journal of Power Sources*, 2011. 196(15): 6465–6472; (e) Cheah, Y.L., V. Aravindan and Madhavi, Electrochemical Lithium Insertion Behavior of Combustion Synthesized V_2O_5 Cathodes for Lithium-Ion Batteries. *Journal of The Electrochemical Society*, 2012. **159**(3): pp. A273–A280; (f) Cheah, Y.L., R. von Hagen, V. Aravindan, R. Fiz, S. Mathur and S. Madhavi, High-rate and elevated temperature performance of electrospun V_2O_5 nanofibers carbon-coated by plasma enhanced chemical vapour deposition. *Nano Energy*, 2013. **2**(1): pp. 57–64; (g) Huang, X., X. Rui, H.H. Hng and Q. Yan, Vanadium Pentoxide-Based Cathode Materials for Lithium-Ion Batteries: Morphology Control,

Carbon Hybridization, and Cation Doping. *Particle & Particle Systems Characterization*, 2015. **32**(3): pp. 276–294.

63. (a) Maher, K. and R. Yazami, A study of lithium ion batteries cycle aging by thermodynamics techniques. *Journal of Power Sources*, 2014. **247**(0): pp. 527–533; (b) Madhavi, S., G.V. Subba Rao, B.V.R. Chowdari and S.F.Y. Li, Synthesis and Cathodic Properties of $LiCo_{1-y}Rh_yO_2$ ($0 \leq y \leq 0.2$) and $LiRhO_2$. *Journal of The Electrochemical Society*, 2001. **148**(11), A1279-A1286.
64. Rui, X., X. Zhao, Z. Lu, H. Tan, D. Sim, H.H. Hng, R. Yazami, T.M. Lim and Q. Yan, Olivine-Type Nanosheets for Lithium Ion Battery Cathodes. *ACS Nano*, 2013. **7**(6): pp. 5637–5646.
65. Aravindan, V., W. Chuiling, Madhavi, S., High power lithium-ion hybrid electrochemical capacitors using spinel $LiCrTiO_4$ as insertion electrode. *Journal of Materials Chemistry*, 2012. **22**(31): pp. 16026–16031.
66. Aravindan, V., M.V. Reddy, S. Madhavi, S.G. Mhaisalkar, G.V. Subba Rao and B.V.R. Chowdari, Hybrid supercapacitor with nano-TiP_2O_7 as intercalation electrode. *Journal of Power Sources*, 2011. **196**(20): pp. 8850–8854.
67. (a) Aravindan, V., M. Ulaganathan, W.C. Ling, Madhavi, S., Fabrication of New 2.4V Lithium-Ion Cell with Carbon-Coated $LiTi_2(PO_4)_3$ as the Cathode. *ChemElectroChem*, 2015. **2**(2): pp. 231–235; (b) Arun, N., V. Aravindan, W.C. Ling and S. Madhavi, Carbon coated $LiTi_2(PO_4)_3$ as new insertion anode for aqueous Na-ion batteries. *Journal of Alloys and Compounds*, 2014. **603**(0): pp. 48–51; (c) Aravindan, V., W. Chuiling, Madhavi, S., Electrochemical performance of NASICON type carbon coated $LiTi_2(PO_4)_3$ with a spinel $LiMn_2O_4$ cathode. *RSC Advances*, 2012. **2**(19): pp. 7534–7539.
68. Aravindan, V., J. Sundaramurthy, A. Jain, P.S. Kumar, W.C. Ling, S. Ramakrishna, M.P. Srinivasan and S. Madhavi, Unveiling $TiNb_2O_7$ as an Insertion Anode for Lithium Ion Capacitors with High Energy and Power Density. *ChemSusChem*, 2014. **7**(7): pp. 1858–1863.
69. Wang, L. P., L. Yu, R. Satish, J. Zhu, Q. Yan, M. Srinivasan and Z. Xu, High-performance hybrid electrochemical capacitor with binder-free Nb_2O_5 graphene. *RSC Advances*, 2014. **4**(70): pp. 37389–37394.
70. (a) Aravindan, V., J. Gnanaraj, Y.-S. Lee and S. Madhavi, Insertion-Type Electrodes for Nonaqueous Li-Ion Capacitors. *Chemical Reviews*, 2014. **114**(23): pp. 11619–11635; (b) Jain, A., V. Aravindan, S. Jayaraman, P.S. Kumar, R. Balasubramanian, S. Ramakrishna, S. Madhavi and M. Srinivasan, Activated carbons derived from coconut shells as high energy density cathode material for Li-ion capacitors. *Scientific reports* 2013, 3. 3002
71. (a) Aravindan, V., Y.-S. Lee, R. Yazami and S. Madhavi, TiO_2 polymorphs in 'rocking-chair' Li-ion batteries. *Materials Today*, 2015. **18**(6): pp. 345–351; (b) Nguyen, L. H., V. Aravindan, S.A. Kulkarni, F. Yanan, R.R. Prabhakar, S.K. Batabyal and S. Madhavi, Self-Assembled Ultrathin Anatase TiO_2 Nanosheets with Reactive (001) Facets for Highly Enhanced Reversible Li Storage. *ChemElectroChem*, 2014. **1**(3): pp. 539–543; (c) Sundaramurthy, J., V. Aravindan, Suresh P. Kumar, S. Madhavi and S. Ramakrishna, Electrospun $TiO_2-\delta$ Nanofibers as Insertion Anode for Li-Ion Battery Applications. *The Journal of Physical Chemistry C*, 2014. **118**(30): pp. 16776–16781; (d) Zhang, X.,

P. Suresh Kumar, V. Aravindan, H.H. Liu, J. Sundaramurthy, S.G. Mhaisalkar, H.M. Duong, S. Ramakrishna and S. Madhavi, Electrospun TiO_2–Graphene Composite Nanofibers as a Highly Durable Insertion Anode for Lithium Ion Batteries. *The Journal of Physical Chemistry C*, 2012. **116**(28): pp. 14780–14788; (e) Jayaraman, S., V. Aravindan, P. Suresh Kumar, W.C. Ling, S. Ramakrishna and S. Madhavi, Synthesis of porous $LiMn_2O_4$ hollow nanofibers by electrospinning with extraordinary lithium storage properties. *Chemical Communications*, 2013. **49**(59): pp. 6677–6679; (f) Arun, N., V. Aravindan, S. Jayaraman, N. Shubha, W.C. Ling, S. Ramakrishna and S. Madhavi, Exceptional performance of a high voltage spinel $LiNi_{0.5}Mn_{1.5}O_4$ cathode in all one dimensional architectures with an anatase TiO_2 anode by electrospinning. *Nanoscale*, 2014. **6**(15): pp. 8926–8934; (g) Zhang, X., V. Aravindan, P.S. Kumar, H. Liu, J. Sundaramurthy, S. Ramakrishna and S. Madhavi, Synthesis of TiO_2 hollow nanofibers by co-axial electrospinning and its superior lithium storage capability in full-cell assembly with olivine phosphate. *Nanoscale*, 2013. **5**(13): pp. 5973–5980; (h) Suresh Kumar, P., V. Aravindan, J. Sundaramurthy, V. Thavasi, S.G. Mhaisalkar, S. Ramakrishna and S. Madhavi, High performance lithium-ion cells using one dimensional electrospun TiO_2 nanofibers with spinel cathode. *RSC Advances*, 2012. **2**(21): pp. 7983–7987; (i) Aravindan, V., Y.S. Lee, Madhavi, S., Research Progress on Negative Electrodes for Practical Li-Ion Batteries: Beyond Carbonaceous Anodes. *Advanced Energy Materials*, 2015, 5(13): pp. 1402225 (j) Tang, Y., Y. Zhang, J. Deng, J. Wei, H.L. Tam, B.K. Chandran, Z. Dong, Z. Chen and X. Chen, Mechanical Force Driven Growth of Elongated Bending TiO_2-based Nanotubular Materials for Ultrafast Rechargeable Lithium Ion Batteries. *Advanced Materials*, 2014. **26**(35): pp. 6111–6118.

72. (a) Jiang, J., Y. Li, J. Liu, X. Huang, C. Yuan and X.W. Lou, Recent Advances in Metal Oxide-based Electrode Architecture Design for Electrochemical Energy Storage. *Advanced Materials*, 2012. **24**(38): pp. 5166–5180; (b) Rui, X., H. Tan and Q. Yan, Nanostructured metal sulfides for energy storage. *Nanoscale*, 2014. **6**(17): pp. 9889–9924.

73. Arun, N., V. Aravindan, S. Jayaraman, N. Shubha, W.C. Ling, S. Ramakrishna, Madhavi, S., Exceptional performance of high voltage spinel $LiNi_{0.5}Mn_{1.5}O_4$ cathode in all one dimensional architecture with anatase TiO_2 anode by electrospinning. *Nanoscale*, 2014. **6**: pp. 8926–8934.

74. Brutti, S., V. Gentili, P. Reale, L. Carbone and S. Panero, Mitigation of the irreversible capacity and electrolyte decomposition in a $LiNi_{0.5}Mn_{1.5}O_4$/nano-TiO_2 Li-ion battery. *Journal of Power Sources*, 2011. **196**(22): pp. 9792–9799.

75. (a) Lu, X., Z. Jian, Z. Fang, L. Gu, Y.-S. Hu, W. Chen, Z. Wang and L. Chen, Atomic-scale investigation on lithium storage mechanism in $TiNb_2O_7$. *Energy & Environmental Science* 2011, 4 (8), 2638–2644; (b) Tang, K., X. Mu, P.A. van Aken, Y. Yu and J. Maier, "Nano-Pearl-String" $TiNb_2O_7$ as Anodes for Rechargeable Lithium Batteries. *Advanced Energy Materials*, 2013. **3**(1): pp. 49–53; (c) Saritha, D. and U.V. Varadaraju, Studies on electrochemical lithium insertion in isostructural titanium niobate and tantalate phases with shear ReO_3 structure. *Materials Research Bulletin*, 2013. **48**(7): pp. 2702–2706.

76. Kim, H., M.-Y. Cho, M.-H. Kim, K.-Y. Park, H. Gwon, Y. Lee, K.C. Roh and K. Kang, A Novel High-Energy Hybrid Supercapacitor with an Anatase TiO_2–Reduced Graphene Oxide Anode and an Activated Carbon Cathode. *Advanced Energy Materials*, 2013. **3**(11): pp. 1500–1506.
77. Karthikeyan, K., S. Amaresh, S.N. Lee, X. Sun, V. Aravindan, Y.-G. Lee and Y. S. Lee, Construction of High Energy Density Supercapacitors from Pine Cone Derived High Surface Area Carbons. *ChemSusChem*, 2014. **7**(5): pp. 1435–1442.
78. Vargas C, O.A., A. Caballero and J. Morales, Can the performance of graphene nanosheets for lithium storage in Li-ion batteries be predicted? *Nanoscale*, 2012. **4**(6): pp. 2083–2092.

50 Years of Biomaterials Research in Singapore

Subbu Venkatraman*, Swee Hin Teoh[†]
and Ali Miserez[‡]

Overview of Publication Activity

Biomaterials research had been pursued at a very low level in Singapore in the period 1980–1995. Most of the early efforts were focused on bone replacement materials, which was not surprising since the earliest recognized biomaterials of any practical importance were developed for dental and hip joint replacement by Burnley in the early sixties. These pockets of biomaterials research in the early period (1980–1995) were at the National University of Singapore and to a lesser extent, at the Nanyang Technological University.

In Singapore, medical materials research may have begun with a company called Pacific Biomedical, founded by the late heart transplant surgeon, Dr Victor Chang. This company started research on heart valve materials (both tissue and synthetic materials were developed) and the first publication "Delrin as an occluder material" was published in ASAIO *Journal for Artificial Internal Organs*, **36** (No. 3), M417–M421, 1990. However after the tragic death of Dr Victor Chang, research activity on heart valve material also came to a close from 1991 until it was picked up again at Nanyang Technological University by Prof Tony Yeo, as described later.

We based our historical account partly on a database (Scopus). Searching for keywords "biomaterials" or "drug delivery" or "tissue engineering" or "nanomedicine"

---

* Professor and Chair, School of Materials Science and Engineering, Nanyang Technological University.
[†] Professor and Chair, School of Chemical and Biomedical Engineering, Nanyang Technological University.
[‡] Associate Professor, School of Materials Science and Engineering, Nanyang Technological University.

or "biomimetic materials" in Title/Abstract/Keywords and "Singapore" as Affiliation Country yielded the desired results (3420 documents). Although not comprehensive in that this approach does not capture ALL of the biomaterials-related activity in Singapore, it does capture the most significant efforts. A separate search of the patent database will be discussed later on in this chapter.

The quantity of biomaterials publication by Singaporean institutions was scanty up to 1995; it picked up somewhat in 1997 and 1998, with the start of activity on hydroxy apatite materials, primarily related to its processing. The number of papers really mushroomed after 2000, when 23 papers were mentioned with Hutmacher (who did his PhD with Prof Teo Swee Hin at the National University of Singapore) as one of the principal authors; one seminal paper among these had attracted over 2000 citations by 2015. The number of papers more than doubled in 2001 (52) reaching triple figures (101) by 2003. Activity in those years saw a qualitative change from hard tissue engineering/replacement to soft tissue engineering and to drug delivery applications, using biodegradable materials such poly (L-lactide) and its copolymers.

From 2003 onwards, the activity was quite frenetic, with publications reaching a record high of 385 in 2013 (more than one per day!) with an apparent levelling off at around 300 papers per year prior to and after 2013. This result speaks of maturity in the field, and of a critical mass being reached, with more novel applications, such as nanomedicine making an appearance around 2010. The nascent discipline of biomimetic or bio-inspired materials also reared its head in Singapore around 2010 with some significant publications from Nanyang Technological University.

During this period the National University of Singapore was the predominant affiliation of the researchers with 1779 of the 3420 publications; NTU came second with 907 papers, and the Yong Loo Lin School of Medicine was third with 276. The more recently-formed Institute of Bioengineering and Nanotechnology (IBN) with 225 and the Institute of Materials Research and Engineering (IMRE, founded in 1997) with 174, continue to be substantial contributors.

The journal *Biomaterials* was the journal of choice for the majority of contributors (298) with the *Journal of Biomedical Materials* (JBMR, parts A & B), *Journal of Controlled Release, Biomacromolecules, Acta Biomaterialia* and *International Journal of Pharmaceutics* all being popular with more than 50 publications each.

Overview of Patent Activity

The earliest biomaterials-related patent from Singapore (US patent database, 1976-present) appeared to be that of Prof Teoh Swee Hin (other inventors were Dr R. Thampuran, Dr James Goh and Dr Winston Seah) in 1996. The patent disclosed a titanium-graphite composite with improved wear characteristics as applied to hard tissue replacement. A second granted US patent appeared in 1998, assigned to NUS, with the first inventor being Eugene Khor, and was related to a chitin preparation to be used as a reinforcing filler, also in hard tissue replacement. It was clear that the early patent activity in biomaterials in Singapore involved hard tissue replacement applications using new biomaterials. This was reflective also of the world-wide trend in biomaterials activity, following Charnley's work on bone cements (polymethyl methacrylate) and on metal-on-polymer articulation in the 1960's.

The focus of the patent activity shifted from hard tissue replacement to drug delivery in the early 2000's. One of the earliest granted patents (filed 2003, granted 2007) was assigned to IMRE, and described a hydrogel made from cyclodextrin and an amphiphilic copolymer for the controlled delivery of bioactive molecules. The inventors were Jun Li, Xu Li, Xiping Ni and Kam Leong. There was no indication of a license or commercialization of this patent. A patent assigned to a company called Genecure in Singapore was filed in 2002 and granted in 2008; this one also deals with the use of cyclodextrin to encapsulate cholesterol, and incorporated genetic material in a cationic polymer for gene transfection.

The next wave of patents in drug delivery would relate to micellar systems and block copolymers of various kinds. The activity was substantial, at A*STAR Institutes, such as IMRE and IBN, and at Nanyang Technological University. The chief inventors were Dr Yang Yi Yan at IBN and Profs Subbu Venkatraman and Freddy Boey at NTU. There was, again, no evidence that any of these drug delivery patents had led to a commercial product; the only one that seemed to have advanced into human trials was a nanoliposome patent from NTU that was used for the prolonged treatment of glaucoma with a single sub-conjuctival injection. A company had been formed (Peregrine Ophthalmic) to take the product through clinical trials in the US.

Another biomaterials-related activity was in the area of degradable implants. A search of the Espacenet database (European patent association) revealed 12 patents in this area, which included stents (coronary, tracheal) as well ocular implants of various types. Some of these patents have been licensed to start-up companies, of which Amaranth Medical (Founded by NTU researchers Venkatraman and Boey) is one example. There was evidence that patent activity in this area led to the translation of ideas.

Another example of a translated patent was a tissue retractor patent, from Nanyang Technological University, with a priority date of 2006-03-31. The concept had moved through trials to USFDA approval and has realized commercial sales. Another biomaterials-related patent was by Yeo Joon Hock (Tony Yeo) of NTU, which described a prosthetic heart valve (1997) assigned to a company (CardioTech); a related patent was one on a valve mold and prosthesis, with a 2006 priority date. The latter concept has since been tested in human trials in India with some success.

In the area of soft tissue replacement, most of the activity has been from NUS, from the laboratory of Prof Ramakrishna, with an emphasis on electrospun fibers as the basis of the constructs. An example was the concept of medical guide tubes (2004-06-04), which described microbraided polymeric fibers (biodegradable) forming a conduit for tissue regeneration, specifically guided nerve tissue regeneration.

Another similar example (2009-08-20) was the use of electrospinning to deposit aligned fibers on the surface of a metallic stent, to guide cellular deposition and alignment. Two spin-offs based on electrospinning, ElectrospunRA and ElectrospinTech, have been announced, concentrating on electrospinning equipment. A third spin-off from NUS, BioMers, was also based at least partially on electrospun fibers forming part of a composite material used in the product SimpliClear, which are transparent or translucent tooth braces used by dentists.

The area of biomaterials for hard tissue replacement including traumatic skull repair had also seen patent activity leading to translation: for example, there were the patents assigned to Ostepore International with the first-named inventor being Prof Teoh Swee Hin, formerly of NUS and currently at NTU. The company has since implanted the device in more than 1000 patients worldwide with FDA 510K and CE mark approval.

More details of patent activity is covered under each sub-category of biomaterials, below.

We will now review accomplishments in each sub-area of biomaterials research and development, with a separate section on the commercialization of research in these fields.

- Drug Delivery
- Medical Devices
- Hard Tissue Engineering

Drug Delivery

A search in the Web of Science revealed some early and pioneering work at the Department of Pharmacy, Singapore (Prof Lucy Wan) on drug delivery. This early work would relate to microspheres from natural polymers such as alginate and chitosan, and concentrated on process variables and their effects on drug incorporation and release. There was sporadic activity in drug delivery at NUS, but not many patent filings in the years 1990–2000. Interestingly, there were reports of clinical trials of externally-developed drug delivery systems, particularly in ocular drug delivery, at the Singapore National Eye Centre, an early pioneer in drug delivery research.

Significant research in the area of materials for drug delivery began in Singapore around the year 2000. Prof Feng at NUS published one of the earliest papers in this field in 2000, on the use of biodegradable polymer nanospheres for solubilizing paclitaxel and controlling its release. *In vitro* the release of paclitaxel was sustained over 6 months!

Prof Feng over the next few years built on this work by exploring the use of vitamin E as an aid in the preparation of PLGA nanospheres and published substantially in related fields. In particular, he also developed degradable micellar systems based on copolymer of PLA with PEG as another type of nanocarrier for paclitaxel.

A number of publications (about 100 in all) attested to the extension of the work on PLGA nanoparticles by Dr Feng. Most of the applications were aimed at cancer chemotherapy, with the major benefits of the nanocarrier approach being passive targetting of tumor tissue; antibody-conjugation was also proposed to enhance "active" targetting of breast tumors, and TPGS (d-alpha-tocophery polyethylene glycol 1000 succinate)-modified PLGA nanoparticles have been studied for delivering chemotherapeutic agents to the brain. This body of work represented the earliest Singapore efforts in the nascent field of Nanomedicine.

It should be noted that the biodegradable polymers themselves were not "discovered" in Singapore; rather, they had been synthesized elsewhere and were commercially available by the late 1990's.

Another extensive body of work related to drug delivery emerged from the Institute of Materials Research and Engineering (IMRE), with some of the research migrating to the Institute for Bioengineering & Nanotechnology when it was formed in 2004. Dr Yang Yi Yan started work on microspheres for drug delivery in the early 2000's, and subsequently into nanoparticles around 2003. The first nanoparticle type she studied was a self-assembling micellar system comprising of end-linked cholesterol to poly acrylamide copolymers to generate temperature-sensitive ("triggered") drug delivery systems. Additionally, there were papers published by Dr Yang on pH-triggered nanoparticle systems which also had a temperature-sensitive motif; essentially these were based on poly acrylamides and 10-undecenoic acid as the pH-sensitive moiety. Both the temperature- and pH-sensitivities were fine-tuned to address preferential release of potent anti-cancer drugs such as doxorubicin, to take advantage of the differential temperature and pH-profiles in solid tumors. A number of high-quality papers attested to the novelty of this work in the mid-2000's, along with a few granted patents on micellar compositions as well as temperature- and pH-sensitive polymeric systems.

Dr Yang continued along the same trajectory of developing unique core-shell nanoparticles for drug/protein/gene delivery; the culmination of this work was a highly-cited publication in Nature Materials (2006) on a plasmid-DNA delivery system made from cationic core-shell nanoparticles that were formed by self-assembly of a biodegradable amphiphilic copolymer; this system was touted to have cellular and nuclear penetration capabilities, as well as the capacity to co-deliver drugs with the plasmid DNA. There was another landmark paper in 2009, this time in Nature Nanotechnology, describing a novel self-assembled cationic peptide system acting as an anti-microbial agent. This was an elegant and complex system derived from 4 moieties: cholesterol, glycine, arginine and the cell-penetrating peptide, TAT; the self-assembled largely spherical nanostructures were active against a wide range of bacteria.

Poly{(N-methyldietheneamine sebacate)-co-[(cholesteryl oxocarbonylamido ethyl) methyl bis(ethylene) ammonium bromide] sebacate} (P(MDS-co-CES))

A novel copolymer synthesized by Dr Yang Yi Yan's group at IBN. This copolymer has the capacity to load both plasmid DNA and a lipophilic drug, for co-delivery to tumor sites. The delivery system was a self-assembled structure (From *Nature Materials* (Letters), volume 5, Issue 10, pp. 791–796 (2006)) that was fully biodegradable into components that are harmless.

There was a US patent application on cationic peptides in 2009, and several other patents based on new biomaterials, as noted below. A search of the USPTO did not yield any application for the DNA vector concept.

Most of the patent activity for Dr Yang centred around her collaboration with IBN polymer scientists, Drs Hedrick and Coady. A number of patents relating to polycarbonates of various types had been granted in the years 2003–2015. Most of these were composition-of-matter patents, and to date, do not appear to have been licensed out for any application.

Overall, the drug delivery work at IMRE and at IBN centres had been around micro- and nanoparticles, as well as on hydrogels. A number of new polymer and peptide-based systems had emerged out of this work, with a focus on applications where cellular penetration/disruption and nuclear delivery was important. With the impressive pre-clinical work reported on these systems, it is perhaps only a matter of time before translation to the clinic happens.

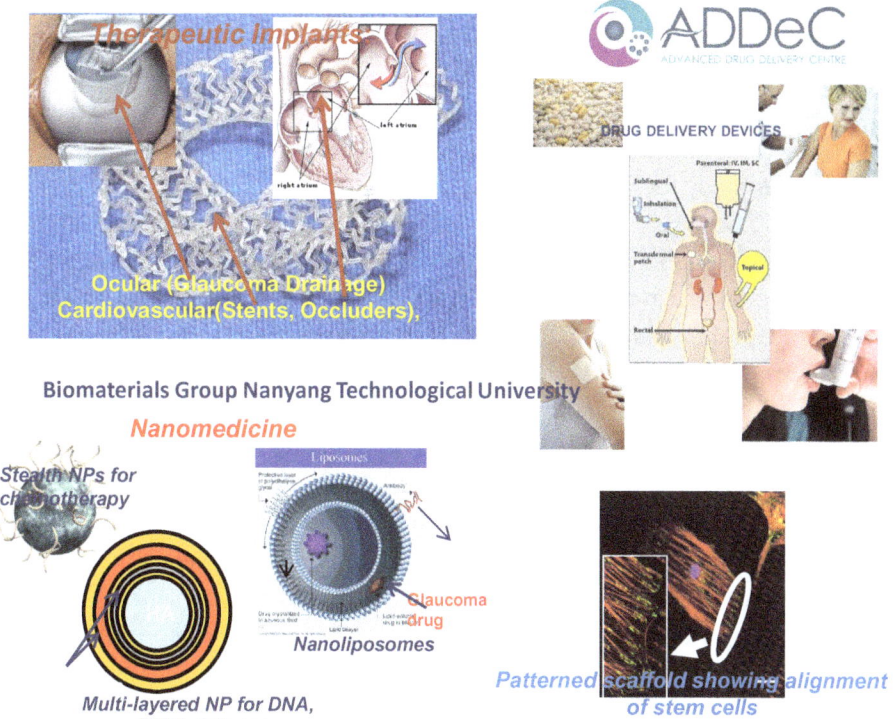

Different areas of biomaterials research at the Biomaterials group of Professors Boey, Venkatraman and others at Nanyang Technological University. The emphasis was on translatable research involving medical devices and drug delivery.

In the same time period (2000 onwards) there was considerable translational work on biomaterials and drug delivery at NTU, primarily in the group headed by Professors Boey and Venkatraman. The work led to commercializable medical devices as well as drug delivery products (pharmaceuticals). Let us discuss the drug delivery aspects first.

The emphasis of the work in the NTU biomaterials group was on sustaining the release of the bioactive molecule (drug, protein, peptide) and localizing its delivery near the tissue of interest. Early on, the group developed a variety of platforms, from microparticles to injectable, *in situ* gelling systems as well as hydrogels. Another platform is a coating (fully-degradable or biostable) on medical devices that can deliver a drug locally. The injectable *in situ* gelling system was a sub-cutaneous injection that was useful for peptide/protein delivery, with the group's work focusing on minimizing the unwanted initial "burst" release. The work on drug-eluting stents was developed with a coater that was adapted for coating onto metallic and polymeric stents. The group worked on a dual-drug eluting stent that co-delivered an anti-thrombotic and an anti-proliferative from the same coating, to yield reduced restenosis in pig studies. This work attracted the attention of a Singaporean cardiologist at Mayo Clinic

(Dr Horng Chen) who was developing some novel chimeric natriuretic peptides for cardiovascular disease. Together, the teams developed *in situ* gelling systems for the sustained delivery of these peptides which were previously only given via intravenous injection. A number of publications arose out of this work, as well as a granted patent. Pre-clinical work was successful in demonstrating efficacy of the delivery system, and the system is ready for human trials.

It should be noted here that concurrent with the work on coated stents, the group also developed fully-degradable implants, including coronary and tracheal stents; more on this in the section on medical devices.

A concept that has proceeded much faster to human trials is a nanomedicine concept for ocular therapy. While the work at IMRE and at NUS in nanomedicine concentrated on nanoparticles as used for targeted or triggered delivery of cancer therapeutics, the NTU group, along with collaborators from the Singapore Eye Research Institute (SERI), principally Dr Tina Wong and Dr Donald Tan, started to apply sustained delivery nanomedicine towards the treatment of ocular disease, in 2009. Glaucoma is a particularly pervasive ocular disease that can slowly lead to blindness if not arrested with chronic therapy. Many patients become blind because they do not apply their treatment (which currently is a daily eye drop with drug in it), consistently. The group developed a localized delivery system using nanoliposomes that eventually went into human trials. In the human trial, it was found that a single sub-conjuctval injection of the nanomedicine exerted its anti-glaucoma effects for up to 6 months, which is a huge advance in sustained-release nanomedicine. This product was being further developed by a spin-off company, Peregrine Ophthalmic.

PEREGRINE Ophthalmic

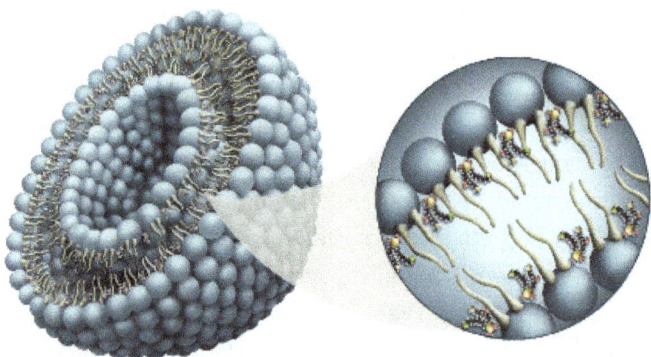

Liposomal latanoprost was an extended release drug delivery solution for glaucoma and was a ground breaking development in nanomedicine. Liposomal latanoprost had completed a small pilot Phase 2a trial. The trial period was 3 months, and all the patients who were previously on prostaglandin medication still showed a clinically significant ocular hypotensive effect at 3 months from injection. http://peregrineophthalmic.com/products/index.html

The glaucoma work in Nanomedicine was recognized with a President's Technology Award in 2014, to the three principal investigators, Profs Venkatraman and Boey, and Dr Tina Wong. This may have been the first President's technology medal awarded for biomaterials research. It also led to the founding of a new institute at NTU, the NTU-Northwestern Institute for Nanomedicine (NNIN), which was co-funded by NTU and Northwestern University's International Institute for Nanotechnology, and will focus on non-cancer applications of nanomedicine.

Medical Devices:

Research into biomaterials that formed a significant and critical part of a medical device, had been much less significant in Singapore until the early 2000's. Most of the pre-2000 work had been focussed on orthopaedic materials and semi-artificial heart valves. Noteworthy among these was the work of collaborators Philip Cheang and Michael Khor at NTU: their work concentrated on hydroxy apatite and derivatives, in particular, the development of coatings made from these materials onto

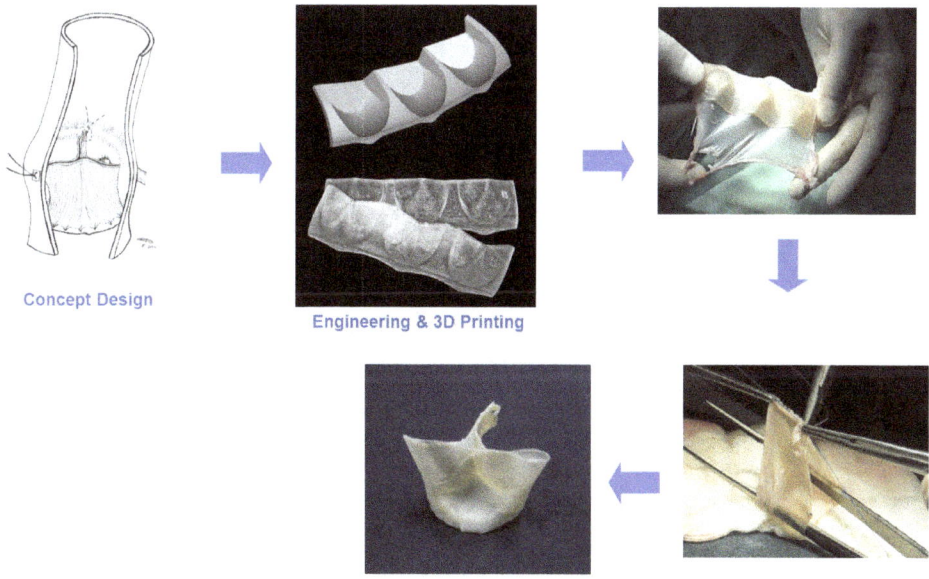

Prof Tony Yeo's patented heart valves, fashioned using a proprietary mold design. The mold enabled heart valve construction using natural biomaterials. The single point attached commissure or SPAC stentless valve required only a single stitch at each commissure with circumferential stitches at the bottom of the valve. This made implantation of the valve simpler and faster

metal or ceramic implants, for the purpose of "compatibilization" and facilitating healing of the implanted device.

Another focus at NTU was on bio-artificial aortic valves, needed for patients suffering from congenital heart valve disease. Led by Dr Tony Yeo Joon Hock, this group developed a stentless valve, templated on a patented mold, that overcame the problems of clotting and hemodynamic perturbation of other valves. Fashioned out of porcine pericardium, this valve could be now implanted relatively simply, and had been successfully tested in humans.

On the other hand, the medical device development work that put Singapore on the world map was done at a company rather than in academia. The company Biosensors International was founded by Yoh Chie Lu in 1995, as a supplier of critical care products. Initially the company focussed on catheter manufacturing, but shifted to drug-eluting stents (DES) around the year 2000. From a biomaterials standpoint, the company developed a fully-degradable coating for drug-eluting coronary stents, called BioMatrix, that incorporated a proprietary anti-restenotic drug called Biolimus. BioMatrix was the first commercial DES in the world using a biodegradable polymer as a carrier of limus drug in treating coronary artery disease. The carrier polymer is poly (L-lactide) or PLA, degrading slowly over 6–9 months, releasing the lipophilic Biolimus drug over time. The drug release profile and the polymer degradation duration would cover the entire wound healing cycle of the patient.

Till date, over a million of Biosensors' DES have been implanted in patients all over the world with excellent clinical results. Biosensors' DES technology had

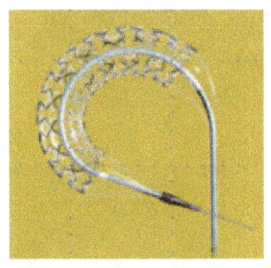
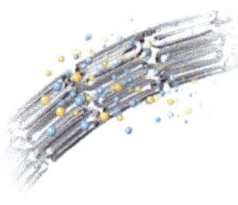

An example of a Biosensors product is the **The Abluminal Biodegradable Polymer DES**. BioMatrix Flex™ is Biosensors' unique abluminal biodegradable polymer DES. Its abluminal coating is absorbed after 6 to 9 months* and turns the DES into a BMS. It combines the proven safety of a DES with an abluminal biodegradable polymer, the proven efficacy of BA9™ and an advanced stent design.

also been successfully proven in a number of clinical trials like LEADERS, STEALTH and SORT OUT.

While Biosensors' work on DES focused on a permanent stent (stainless steel) as the scaffold for keeping arteries open, research at NTU's Biomaterials group (Venkatraman, Boey and Lay Poh Tan) was aimed at replacing the metal scaffold with a fully-degradable material, made from poly (L-lactide) and its copolymers with poly (glycollide); the copolymers are called PLGA's. The group developed a multilayered stent that used the principle of shape memory for deployment of the device in the body. There was growing evidence in the early 2000's that a stent was not really needed in the body after about 6 months, and in fact its permanent presence is sometimes a liability, as it interferes with MRI diagnostics, and may necessitate the lifelong use of anti-clotting drugs in some patients. However, fashioning a self-expanding stent out of biodegradable polymers is a challenge, and took a multi-disciplinary effort to arrive at a proprietary formulation. This patented stent became the basis of a technology that was licensed to venture capitalists (BioOne Capital, Singapore; and Charter Life Sciences, USA) in 2005, with the founding of a company called Amaranth Medical. Now in its 10th year

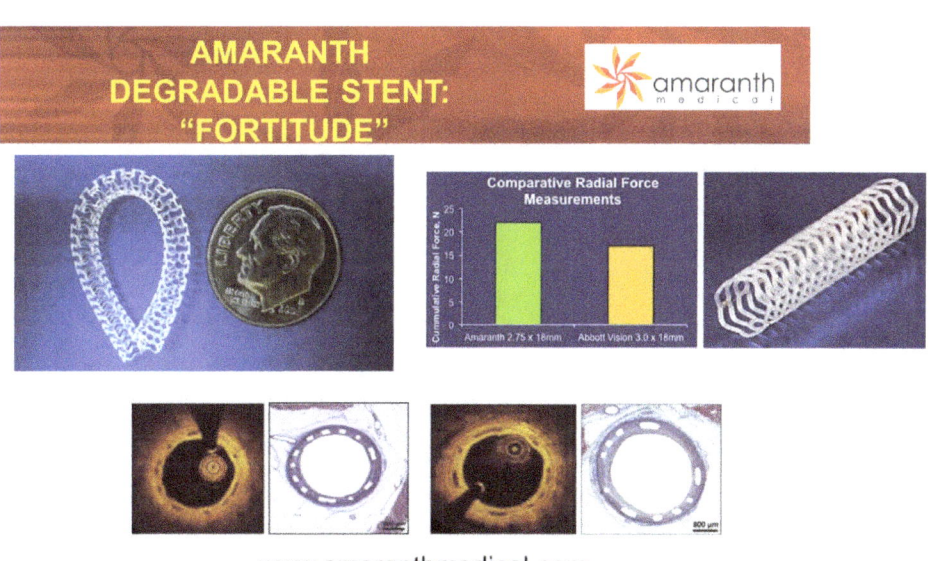

www.amaranthmedical.com

The fully-degradable stent made by Amaranth Medical has already undergone successful human trials, and is set to finish further trials by 2017. The stent, trade-marked Fortitude™, is expected to get FDA approval in 3 years. The company is head-quartered in Mountain View, California, with a branch in Science Park II, Singapore.

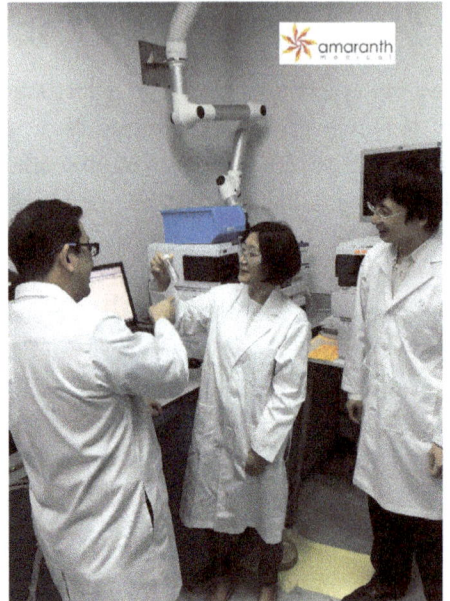

Three of the NTU graduates who work at the Amaranth Medical facility in Science Park II (from left to right):

Dr. Alfred Chia, Dr. Wang Liwei, Dr. Pan Jie.

Fully-Degradable Occluders For:

Patent Foramen Ovale (PFO)
Atrial Septal Defect (ASD)
Patent Ductus Arteriosus (PDA)

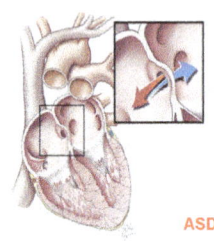

ASD

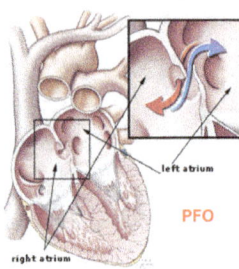

PFO

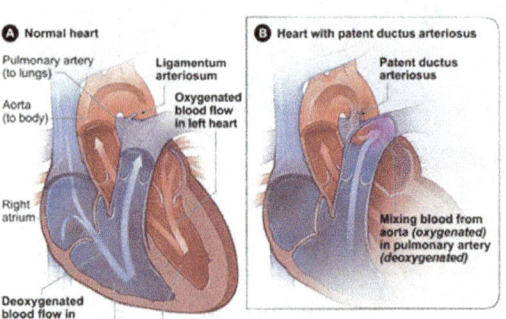

PDA

MARKET ESTIMATE: ASD, VSD, PFO and PDA: ~$250 Million in 2013

Three conditions that would benefit substantially from a fully-degradable plug: PFO, ASD and PDA, most of which occur in the pediatric population. These are currently treated with a metallic (nickel-titanium alloy, Nitinol) occlude that is permanently placed at the defect site, and can prevent access to one side of the heart. (The use of a permanent occluder also leads to problems such as clotting and allergies.) Fully-degradable plugs are a better option, and have been developed and tested at NTU.

of operation, the company's fully-degradable stent, called Fortitude™, has been proven successful in humans and awaits further testing prior to US FDA approval. The company is headquartered in Mountain View, California, with a Singapore branch that does all the material processing for the stent, and also employs several Singapore graduates from NTU. The company's product has attracted the attention of a major stent company, Boston Scientific, and is expected to complete clinical trials by 2017.

Applying a similar approach to sealing heart defects, the same group has developed fully-degradable occluders for atrial septal defects (ASD), patent foramen ovale (PFO) and patent ductus arteriosus (PDA). Such occluders will replace metallic occluders which are hard and permanent, presenting problems of access to the heart as well as problems of clotting. Following successful animal trials, these occluders are ready for human trials.

A fully-degradable ureteric stent (Venkatraman, Boey, Huang YingYing and Chia Sing Joo of TTSH) as well as a fully-degradable tracheal stent (Venkatraman, Anthony Ng and Lynne Lim of NUH) have also been developed and tested successfully in animals. Patents relating to these stents, both of which have a drug-eluting function as well, have been filed.

Hard Tissue Engineering

Hard tissue engineering research in Singapore started in the mid 1990's, primarily with NUS researchers looking at fatigue and wear of biomaterials. Fatigue, fracture and wear have been identified as some of the major problems associated with implant loosening, stress-shielding and ultimate implant failure. Although wear is commonly reported in orthopedic applications such as knee and hip joint prostheses, it is also a serious and often fatal experience in mechanical heart valves. Fatigue-wear interaction plays a significant role in the ultimate failure of these medical devices.

Hard tissue engineering research in Singapore gathered momentum in the 1996 with a focus on dental materials — where the motivation was to find suitable biocompatible substitutes for toxic mercury amalgams — and on hip joint materials. Prof Teoh Swee Hin and his team have contributed significantly in this area and books and monographs have been written on this topic. Patents such as a sintered titanium — graphic composite having improved wear resistance and low frictional characteristics (US 5,758,253), and transparent composite membrane (US 6,652,966 B1), were generated.

In the early 2000's, Singapore researchers started to pay attention to the incorporation of cells in biomaterials, in the midst of a world-wide interest in tissue engineering to replace damaged tissue. Tissue engineering has been rightfully

Empowering Natural Tissue Regeneration

Three products from Osteopore International, that can act as scaffolds in craniofacial reconstruction, made using 3-D printing.

heralded as a paradigm-shifting approach to tissue replacement and regeneration, including tissue that normally does not regenerate, such as cardiac and neuronal tissues. This is a highly multidisciplinary field and involves the integration of engineering principles, basic life science, and molecular cell biology. Tissue engineering for a number of tissues or organs such as skin, bladder, cartilage, bone and liver were being researched world-wide, and Singapore joined the bandwagon with generous support from local funding sources.

The work in tissue engineering had also spurred work on biodegradable materials, including polycaprolactone (PCL), which was used as a (non-tissue containing) scaffold for cranial repair. Numerous successful clinical studies has been carried out by Prof SH Teoh and his team. Their work presented successful long term (exceeding 3 years) outcomes in craniofacial and mandible tissue replacement. In 2003 a new company, Osteopore International, was formed to develop and commercialize this PCL-based biomaterial technology. The company was the first to obtain an FDA 510K for scaffolds based on polycaprolactone for use in craniofacial applications. The research in this area has trained several researchers who have since relocated to continue similar activity in hard tissue engineering.

Other Biomaterials Research

A very important contribution to the field of biomaterials is the production of micron- and nanometer-sized fibers via the use of electrospinning. Although such fibers may also be produced by melt-spinning, its importance to biomaterials research is that it enables the production of fibers from both synthetic and natural polymers, and that the process is relatively benign. The extension of this technique to produce core-shell fibers also has been revolutionary in that the core can infact be used to encapsulate sensitive bioactive species, including live cells.

The pioneer in this field in Singapore is Professor Seeram Ramakrishna of the National University of Singapore. At a fairly nascent stage of this technology, Prof Seeram recognized that many living organisms i.e. plants, animals, and humans in Nature are made of nanometer-scale fibers. In other words they are fibers with diameters about one hundredth of the thickness of paper. Many living organisms derive their functional properties from the nanofibers present within them. Along with this recognition came the realization that Nature has solution-based methods for producing such fibers *in situ*, using fairly mild conditions. That in turn led to this ingenious and mild way of producing fibers under the influence of an external electric field, from solutions of natural and synthetic polymers.

Today, nano- and microfibers, produced predominantly by electrospinning, has found a place in biomaterials research. Prof Seeram Ramakrishna's group at NUS, have been prolific in this field, catapulting Singapore to the top of the list in terms of publication numbers (see Table below).

Many biomedical applications have been cited for these fibers, such as tissue engineering scaffolds, and for drug delivery and diagnostics; several patents have been filed. Perhaps the most intriguing application of this technology has been its use in corrective dental braces, made from nanofiber composites that are optically clear: the company is Biomers and the product is SimpliClear. The research has also led to other spin-offs such as Electrospunra, which markets machines intended

Ranking of countries and institutions by their output of research papers on electrospun nanofibrous scaffolds for tissue engineering, 2000 – May 2011.

Papers	Country	Rank	Institution	Papers
657	USA	1	National University of Singapore	144
448	China	2	Songhua University	120
438	S. Korea	3	SUNY Stony Brook	58
161	Singapore	4	Virginia Commonwealth University	56
92	UK	5	Seoul National University	53
80	Italy	6	Chinese Academy of Sciences	42
70	Germany	7	Hungnam National University	35
66	Japan	8	Chulalongkorn University	34
49	Australia	9	Ohio State University	28
39	Thailand	10	University of Pennsylvania	27

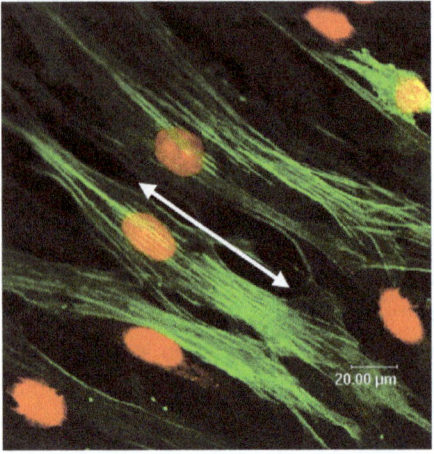

Fabricated muscle cells living on nanofibers, from Prof. Seeram Ramakrishna's laboratory

for the facile electrospinning of polymeric materials; one such equipment is the ES210. All this work has led to Prof Seeram being one of the world's most highly-cited researchers, and more recently to be identified as among the world's most influential scientists, by Thompson Reuters.

The Future of Biomaterials Research

A fledgling area in biomaterials is biomimetics. Simply put, this research evaluates natural materials which have a unique combination of properties, understands the physico-chemical and structural principles responsible for these exceptional properties across multiple length scales (from the genetic and molecular scales all the way up to the meso-scale), and aims at synthetically reproducing them in the laboratory. Furthermore, natural (or biological) materials exhibit two critical features that have sparked interest. First they are deemed biocompatible: in protein-based biomimetic materials for instance, their degradation products are amino acids, which are central elements of the metabolic activity and thus readily recycled by our body. Inflammation and immunogenicity responses are thus unlikely. Second, natural biomaterials are processed under "eco-friendly" conditions, namely at low pressure and temperature, in aqueous conditions, and using natural building blocks. Such bioprocessing, if fully understood and duplicated, could lead to the next-generation of high-performance materials using environmentally-friendly production processes. In Singapore, this effort is spearheaded by researchers at NTU (Prof Ali Miserez) and at NUS (Prof Suresh Valiyaveettil).

Biomimetic Materials Research: Prognosis

Prof. Miserez's laboratory at NTU ("Biological and Biomimetic Materials Laboratory, BBML") has developed a cross-disciplinary approach at the frontiers of Life Sciences and Physical Sciences, blending areas such as protein biochemistry, molecular biology and protein engineering, and multi-scale spectroscopic and mechanical characterization of biological tissues. One key aspect of the team's research strategy is to take advantage of remarkable advances in genomic sequencing technology in the past decade, which has allowed unprecedented access to high-throughput genetic information. For biomimetic materials research, these methods are very powerful to determine the primary structure of proteins that constitute the building blocks of biological materials or that template the growth and self-assembly of inorganic components into high-performance biocomposites. In close collaboration with Dr. Shawn Hoon from the Molecular Engineering Lab (MEL) at A*STAR (MEL

Biological and Biomimetic Materials Laboratory (BBML)

- **Hyper damage-tolerant bioceramics:**
  Mantis shrimp dactyl clubs

- Weaver. et al., **Science**, 2012
- Amini et al, **Nature Communications**, 2014
- Amini et al, **Nature Materials**, 2015

- **Graded structural tissues with no mineralization:**
  Cephalopod beaks

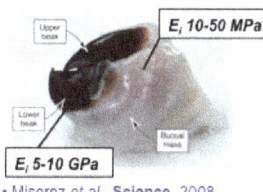

- Miserez et al., **Science**, 2008
- Miserez et al., **J. Biol. Chem.**, 2010
- Tan et al, **Nature Chem. Biol.**, 2015

- **Shock-absorbing biopolymers:**
  Marine snail egg capsules

- Miserez et al., **Nature Materials**, 2009
- Miserez and Guerette, **Chem. Soc. Review**, 2013
- Fu et al., **J. Mat. Chem. B**, 2015

- **Strong, protein-based supramolecular polymers:**
  Cephalopod sucker ring teeth

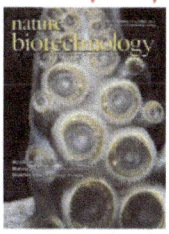

- Guerette et al., **Nature Biotechnology**, 2013
- Guerette et al., **ACS Nano**, 2014
- Miserez et al., **Adv. Materials**, 2009
- Ding et al., **Adv. Materials**, 2015
- Latza et al., **Nat. Comms**. 2015

Model biological materials studied by NTU's BBML. Research includes high-throughput gene and protein sequencing constituting the materials' building blocks, multi-scale structure/property relationships using advanced spectroscopic and nano-mechanical testing, and protein and peptide engineering aimed at duplicating the materials' design principles at the molecular scale.

was founded by Nobel Laureate Dr. Sydney Brenner), Prof. Miserez's group has pioneered the use of Next Generation RNA-sequencing (RNA-seq) technology in their biomaterials research. In parallel, the team also exploits the latest advances in nano-scale mechanical characterization, x-ray scattering methods, or solid-state spectroscopic tools to obtain in-depth understanding of the structure/property relationships of biological materials.

Prof Miserez has concentrated largely on hard tissues from marine invertebrates, and his work has been featured in leading journals such as Science, Nature Biotechnology, Nature Materials, and Advanced Materials. One intriguing application is the development of fracture and wear resistant biomaterials that are more friendly to the host tissue, for instance incorporating mechanical gradients that minimize interfacial stresses, a leading cause of joint implant failures, such as hip and knee implants. However, such materials have more wider-ranging applications as well that will be explored further with the new A*STAR initiative to set up a Centre for Integrative Science, which incorporates Bio-inspired Materials research. This is a truly multi-disciplinary research, in that its successful outcomes depend on contributions from materials scientists, synthetic chemists and biochemists, structural physicists and life scientists. A patent has in fact been filed on the so-called Sucker Ring Protein (SRT) family of proteins, based on mimicking the protein found in the tentacles of squid and cuttlefish; this family of proteins have some unique properties that may well be amenable to a variety of biomedical applications.

Another area of interest to biomaterials researchers is the study of ultra-tough biomineralized materials found in predatory appendages or exoskeletons of marine organisms. In particular, the team has focused their efforts on the ultra impact-tolerant dactyl clubs of the mantis shrimps. High-resolution imaging and spectroscopic measurements have revealed that the dactyl clubs are graded biocomposite consisting of crystalline fluorapatite (FAP), amorphous calcium carbonate, and chitin, with textured FAP the predominant phase near the impact region. The lessons gathered from this understanding provide the design principles for wear-resistant hard tissue implants.

Prof Suresh Valiyaveetil's group at NUS had concentrated on mimicking the ubiquitous egg shell proteins. Using reverse engineering, the team had been able to reconstruct a "synthetic" version of the egg shell in the laboratory. According to the team's findings, almost 98% of the shell material is calcium carbonate. The 2 percent of proteins act as binders for the calcium carbonate. Understanding the structure-property relationships in these egg shells, it was hoped, would lead to bone replacement materials in the future.

The other important classes of biomaterials include silk-based proteins, 2-dimensional graphene- or graphene oxide-based materials as well as nanotubes made mostly from carbon. Such materials are studied in NTU and NUS as well as at IMRE, and are slowly gaining attention for their unique properties with possible biomedical applications. This chapter may well be written over the next 50 years in Singapore.

Selected Readings by Singaporean Biomaterials Researchers:

General Biomaterials

1. Subbu Venkatraman and C.K. Chan, "Biomaterials: Usage in Tissue Engineering and Drug/Gene Delivery", Chapter 3, *Life Sciences, Engineering Applications in Biology*, Tet Fatt Chia, Editor, McGraw Hill Education (Asia), 2002 (BOOK CHAPTER).
2. Teoh Swee Hin, *Engineering Materials for Biomedical Application*, World Scientific, 2004. Book featuring chapters on different biomaterials uses and processing. (BOOK)
3. Seeram Ramakrishna, M Ramalingam, T S Sampath Kumar, and W O Soboyejo, *Biomaterials: A Nano Approach*, CRC Press, 2010 (BOOK).
4. SH Teoh, ZG Tang and GW Hastings, *Thermoplastics in Biomedical applications: Structures, Properties and Processing*, Handbook of Biomaterial Properties, Ed., J Black and GW Hastings, London/UK: Chapman and Hall, 1998, pp. 270–302 (BOOK CHAPTER).
5. Subbu Venkatraman, F Boey and LL Lao. Implanted cardiovascular polymers: Natural, synthetic and bio-inspired, *Progress in Polymer Science*, **33** (9), pp. 853–874 (2008).
6. ZM Huang, YZ Zhang, M Kotaki and S Ramakrishna, A review on polymer nanofibers by electrospinning and their applications in nanocomposites, *Composites Science and Technology*, 63(15), pp. 2223–2253 (2003).

Biomaterials and Drug Delivery

1. P. Y. Teo, W. Cheng, J. L. Hedrick and Y. Y. Yang, Co-Delivery of Drugs and Plasmid DNA for Cancer Therapy, *Advanced Drug Delivery Reviews*, (2015) DOI:10.1016/j.addr.2015.10.014.
2. Subbu Venkatraman, N. Davar, A. Chester and L. Kleiner, *An overview of controlled-release systems*, (Chapter 22). Handbook of Pharmaceutical Controlled-Release Technology, D.Wise, Ed., published by Marcel Dekker in 2000 (BOOK CHAPTER).
3. L Liu, K Xu, H Wang, PKJ Tan, W Fan, Subbu Venkatraman, L Li and YY Yang, Self-assembled cationic peptide nanoparticles as an efficient antimicrobial agent, *Nature Nanotechnology*, **4**, pp. 457–463 (2009).
4. Zhang ZP, Tan SW and Feng SS, Vitamin E TPGS as a molecular biomaterial for drug delivery, *Biomaterials* **33**(19), pp. 4783–4974 (2012).

5. Subbu Venkatraman, K. Keith, Y. Rosen, Y. Huang, T.W.J. Steele, F. Alexis Drug Delivery Systems for Vascular Disease Therapy, in *Polymers for Vascular and Urinogenital Applications*, CRC Press, 2012 (BOOK CHAPTER).
6. X W Ng, R C Mundargi and Subbu Venkatraman, "Nanomedicine: Size-related drug delivery applications, including periodontics and endodontics", *Nanotechnoloy in Endodontics*, Springer, 2014 (BOOK CHAPTER).
7. Subbu Venkatraman and F Boey. Release profiles in drug-eluting stents: Issues and uncertainties, *Journal of Controlled Release*, **120**(3), pp. 149–160 (2007).
8. Jayaganesh Natarajan, Ng Xu Wen, Chandra Nugraha and Subbu Venkatraman, Sustained Release from Nanocarriers — a Review, invited article, *Journal of Controlled Release*, 193, pp. 122–139 (2014).

Biomaterials and Devices/Tissue Engineering

1. M. Ramalingam, Z. Haidar, S. Ramakrishna, H. Kobayashi and Y. Haikel, *Integrated Biomaterials in Tissue Engineering*, Scrivener Publishing, 2012.
2. SH Teoh, ZG Tang and S Ramakrishna, Development of thin composite membranes for biomedical applications, *J. Mats. Sci.: Maths Med.*, **10** (1999): 343–352.
3. DW Hutmacher, JCH Goh and SH Teoh, An introduction to biodegradable materials for tissue engineering applications, *Ann. Acad. Med. Singapore*, **30**(2) (2001): 183–191.
4. Y Huang, JF Kong, Subbu Venkatraman, Biomaterials and design in occlusion devices for cardiac defects: a review, *Acta Biomaterialia*, **10**(3), pp. 1088–1101 (2014).
5. Z Ma, M Kotaki, R Inai, S Ramakrishna, Potential of nanofiber matrix as tissue-engineering scaffolds, *Tissue Engineering* **X**(102), pp. 101–109 (2005).
6. Hutmacher DW, Zein I, Teoh SH, Ng KW, Schantz JT, Leahy JC. Design and Fabrication of a 3D Scaffold for Tissue Engineering Bone: in *Synthetic Bioabsorbable Polymers for Implants*, STP 1396, ASTM, pp. 152–167 (2000).
7. SA Irvine, X Yun, Subbu Venkatraman. Anti-platelet and tissue engineering approaches to biomaterial blood compatibilization: how well have these been translated into the clinic? *Drug Delivery and Translational Research*, 2, pp. 384–397 (2012).

Biomimetic Materials

1. Guerette, P.A., Hoon, S., Seow, Y., Wong, F.T., Ho, V.H.B., Raida, M., Masic, A., Demirel, M.C., Abdon, F., Amini, S., Tay G.Z., Ding, D., and Miserez A. (2013). Accelerating the Design of Biomimetic Materials by Integrating RNA-Seq. with Proteomics and Materials Science. *Nature Biotechnology* 31, 908–915.
2. Amini, S., Tadayon, M., Idapalapati, S., and Miserez, A. (2015). The Role of Quasi-Plasticity on the Extreme Contact Damage Tolerance of The Stomatopod Dactyl Club. *Nature Materials* 14, 943–950.

3. Ding, D., Guerette, P.A., Jing, F., Zhang, L., Irvine, S.A., and Miserez, A. (2015). From Soft Self-Healing Gels to Stiff Films in Suckerin-Based Materials Through Modulation of Cross-Link Density and β-Sheet Content. *Advanced Materials* 27, 3953–3961.
4. P. K. Ajikumar, S. Vivekanandan, R. Laxminarayanan, S. D. S. Jois, R. M. Kini and S. Valiyaveettil (2005), Mimicking the function of eggshell matrix proteins: Role of multiplets of charged amino acid residues and self-assembly of peptides in biominerlization, *Angewandte Chem. Int. Edn.* 44, 5476–5479.

7 2D Materials

Andrew T. S. Wee*, Kian Ping Loh†
and Antonio H. Castro Neto‡

Beginnings of 2D Research in Singapore: NUS Surface Science Laboratory (1986–2005) [A Wee]

Surfaces, the two dimensional (2D) termination of a bulk solid, have been notoriously difficult to study. The Physics Nobel Laureate Wolfgang Pauli was quoted as saying:

> *"God made the bulk; surfaces were invented by the devil."*[1]

Indeed the studies of solid surfaces and 2D materials have historically been difficult to study due to their sensitivity to the external environment at the interface, be it a gas, liquid or another solid. Nevertheless, surfaces are critically important in modern technology, such as in the fields of heterogeneous catalysis, optical and electronic devices, protective coatings, adhesion, sensors, energy storage and generation and so on.

The field of surface science encompasses the study of physical and chemical phenomena that occur at the interface of two phases, including solid–liquid interfaces, solid–gas interfaces, solid–vacuum interfaces, and liquid–gas interfaces. It started to gain prominence in the 1970s, when ultrahigh vacuum technology was maturing and it was possible to study surfaces reproducibly in well-controlled conditions at the solid-vacuum interface. Today the fields of surface chemistry and

---

*Vice President, National University of Singapore and President, Singapore National Academy of Science.
†Chief Scientist, Carbon Convergence Technology Laboratory, Singapore.
‡Director, Centre for Advanced 2D Materials, Singapore.

surface physics underpin the foundations of surface engineering applied to the many technologies as mentioned in the preceding paragraph.

The story of surface science and 2D research can be traced to the establishment of the Surface Science Laboratory (SSL) in 1986 at the Department of Physics, National University of Singapore.[2] The first ultrahigh vacuum (UHV) system delivered in 1987, possibly the first in Singapore, was a Vacuum Generators (VG) ESCALAB Mk 2/SIMSLAB (Figure 1), acquired with a S$1.5M Science Council grant. The first Director, physics Professor Tan Kuang Lee, and his chemical engineering collaborators Professors Kang En-Tang and Neoh Koon Gee did pioneering research in the photoelectron spectroscopy of polymers on this instrument, and their achievements were recognised when they won the 1996 National Science Awards (Figure 2).

This author (A Wee) joined NUS in 1990, and participated in the development of the Surface Science Lab from a single UHV system to several state-of-the-art UHV systems today. As one of the earliest major research laboratories in NUS, the Surface Science Laboratory is recognised today as one of the world's leading surface science groups producing numerous international journal publications, patents and international collaborations. The lab also actively engages the local industry, and has organised surface and interface analysis workshops to raise Singapore companies' competency in state-of-the-art materials characterisation techniques (Figure 3).

The laboratory currently houses five ultrahigh vacuum (UHV) systems (as of mid-2015), including the ESCALAB, Omicron low-temperature scanning tunneling microscope (LT-STM), Omicron variable-temperature (VT)-STM, ultraviolet/x-ray photoelectron spectroscopy (UPS/XPS) system, and Cameca IMS 6f Secondary Ion Mass Spectrometry (SIMS) system, along with an array of thin film growth systems, and other characterisation equipment. In 2000, we received an NSTB[3] grant to set up the Surface-Interface-Nanostructure-Science (SINS) beamline and end-station (Figure 4) at the Singapore Synchrotron Light Source (SSLS).[4] The synchrotron source produces highly monochromatic and tunable x-rays and serves as a unique probe to study surfaces and interfaces.

Our journey into graphene and 2D materials research was serendipitous. In the early 2000s, PhD candidate Chen Wei, currently an Associate Professor in NUS Chemistry, was investigating the surface structure of the carbon nanomesh phase formed on top of an annealed silicon carbide surface, SiC(0001). He was first author of a 2005 paper published in the journal *Surface Science* that proposed the structure of the carbon nanomesh, "whereby isolated carbon islands one atomic layer thick assemble to form the nanomesh structure" (Figure 5).[5] These one atomic layer thick carbon islands were actually *graphene* islands, and we

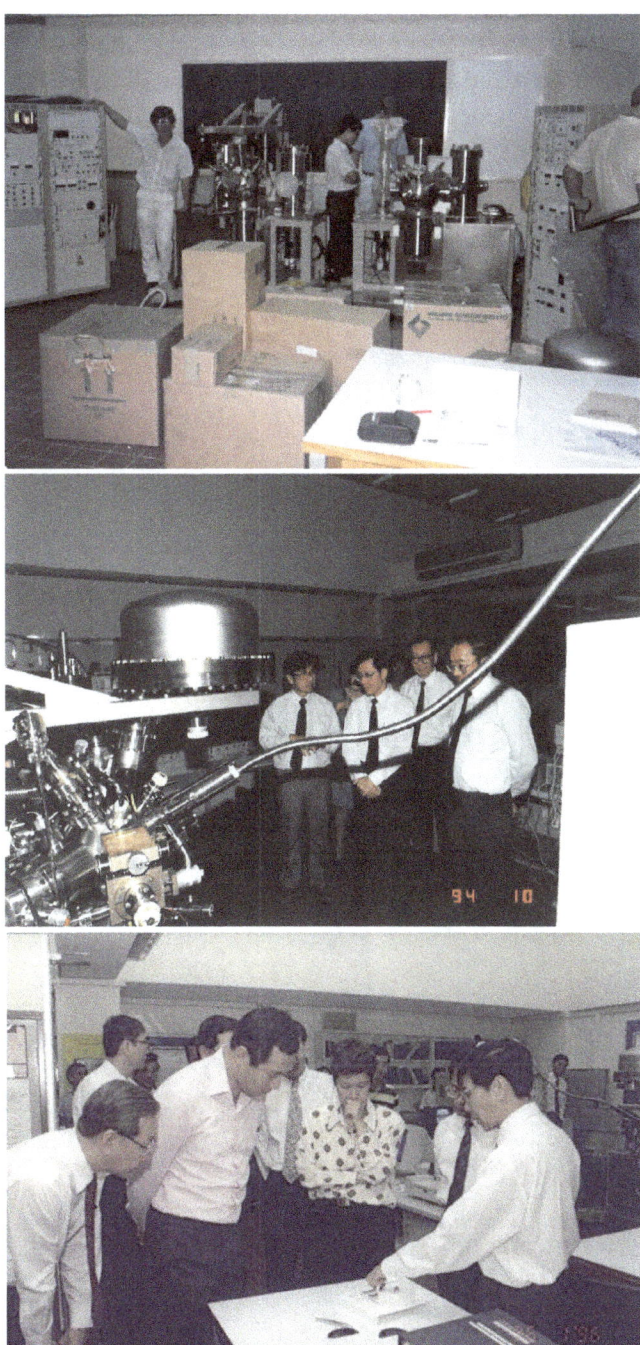

Fig. 1. [Top] Delivery and installation of the VG ESCALAB Mk 2/SIMSLAB at the NUS Surface Science Laboratory; Professor Tan Kuang Lee is standing on the left. [Middle] Prof Tan hosting a 1994 visit of the then Minister for Education, Mr Lee Yock Suan, accompanied by the then Vice Chancellor Professor Lim Pin. [Bottom] Prof Tan hosting a 1996 visit of PM (then DPM) Lee Hsien Loong.

Fig. 2. [Top photo, left to right] Professors Tan Kuang Lee, Kang En-Tang, Neoh Koon Gee posing with the VG ESCALAB Mk 2/SIMSLAB system; and [bottom] winning the 1996 National Science Award. (adapted from *lian he zao bao* article 3 Sep 1996).

Fig. 3. The NUS Faculty of Science newsletter highlighting the Surface and Interface Analysis Workshop organized for industry participants in 1992.

Fig. 4. The Surface-Interface-Nanostructure-Science (SINS) beamline [top], and end-station [bottom] at the Singapore Synchrotron Light Source (SSLS).

had not recognised then the significance of graphene, the reason being that the landmark 2004 paper by Physics Nobel Laureates Novoselov and Geim had only just been published.[6] As a result of the surge of interest in graphene in the years that followed, our paper eventually became a top cited article in *Surface Science* (2005–2010), serendipitous indeed!

Since then the Surface Science Laboratory has published numerous papers on graphene, 2D monolayers, and hybrid organic-inorganic heterostructures. Our LT-STM system has been particularly instrumental in allowing us to visualise atomically resolved images of graphene and adsorbed molecules. Figure 6 shows a photograph of our LT-STM with the growth chamber attached, as well as a typical STM image of epitaxial graphene, which was highlighted in *Nature News* on 25 March 2009.[7] As of 2015, we are currently engaged in STM studies of 2D transition metal dichalcogenide (TMD) monolayers. Unlike graphene, TMDs such as MoS_2 and WSe_2, are semiconductors with tunable direct bandgaps dependent on the number of atomic layers, and have potential electronic and optoelectronic applications.

Fig. 5. [Top] Paper published in the journal *Surface Science* that proposed the structure of the carbon nanomesh "whereby isolated carbon islands one atomic layer thick assemble to form the nanomesh structure" (Chen et al., *Surface Science* **596** (2005) 176); [bottom] Certificate in recognition that this paper was a top cited article in *Surface Science* (2005–2010).

From Diamond to Graphene (1998–2010) [KP Loh]

Loh KP started his research in diamond growth in 1999 in collaboration with A Wee. A home-built hot filament CVD system constructed by MSc student Lin Ting provided

 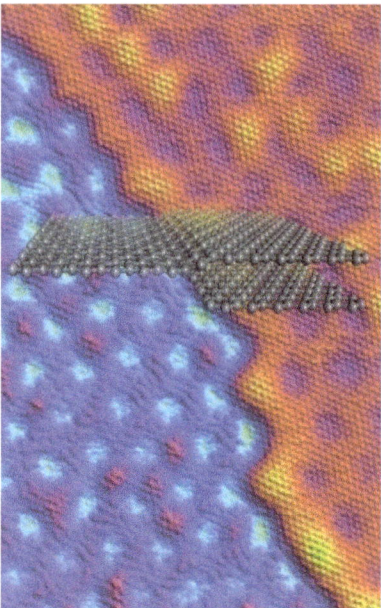

Fig. 6. [Left] Low temperature scanning tunnelling microscope (STM) with growth chamber attached; [right] STM image of epitaxial graphene with model schematic superposed (Nature News 25 March 2009; Nature 458, 390–391 (2009)[8]).

the equipment for their first foray into diamond research.[9] The team was motivated to synthesise nanocrystalline diamond due to the advent of nanotechnology at that time, as well as its potential application as hard tribological coatings. By adding argon into the usual hydrocarbon feedstock used for diamond growth, nanocrystalline diamond could be grown in a narrow composition window. The transition in diamond crystal morphology from well-faceted microcrystalline to nanocrystalline phases as a function of increasing argon concentration was studied. Analysis of the plasma by optical emission spectroscopy revealed a linear correlation between the argon addition and the occurrence of C_2 dimers in the plasma. This work had since received very good citations by the international diamond community.

Recognising that the research focus on diamond at that time was largely engineering driven, Loh KP decided to carve a niche in the surface chemistry of diamond, with a view towards their application as biosensors[10-12] (Figure 7). One perspective he introduced to the community was the notion that a hydrogenated diamond surface can be viewed as the organic analogue of solid hydrocarbon.[13,14] The C-H bonds on the surface of diamond can be activated for C-C coupling in analogy to the mainstream chemistry effort on C-H activation. Diamond surface chemistry research allowed him to establish a unique niche in diamond research, and Loh was invited to be the associate editor of *Diamond*

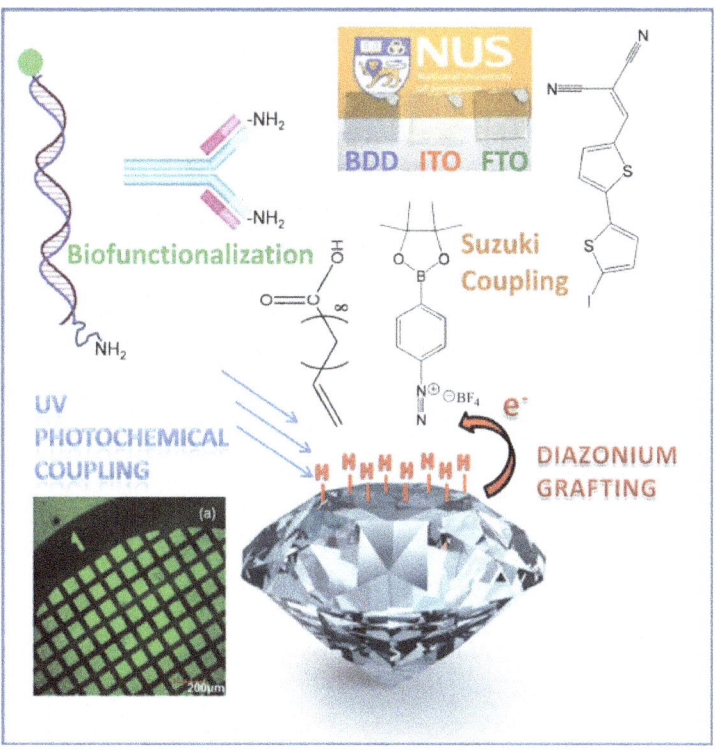

Fig. 7. Surface chemistry of diamond.

and Related Materials (2009–2012). He was also an active member in the advisory panels and scientific boards of international diamond conferences such as the European Diamond Conference series, New Diamond and Nanocarbon and the Hasselt Diamond workshops.

Loh KP started researching the 6H-SiC surface in 2000. Dr Xie Xianning, now the assistant director of NUS Suzhou Research Institute, was his first PhD student and he worked on surface reconstructions on 6H-SiC (0001). In a paper published in *Diamond and Related Materials* in 2001 entitled "Atomic beam etching of carbon superstructures on 6H-SiC (0001) studied by reflection high energy electron diffraction (RHEED)", the various stages during the segregation of carbonaceous super-structures at high temperatures were studied by RHEED.[7,8,15,16] A smooth silicate-terminated root3 x root3 R30 surface could be obtained after a hydrogen-plasma beam treatment of 6H-SiC at 800 degrees C. Annealing the root3 x root3 R30 face to 900 degrees C readily resulted in the segregation of a few layered graphene on the surface, with the basis vectors of the graphite unit cell rotated 30 degrees with respect to the bulk SiC. The importance of this as a means to form single layer graphene had not been recognised at that time. Following Xie's work, Loh KP and A Wee co-supervised a PhD student Chen Wei, who continued to research on the carbon superstructure on 6H-SiC, and this led to further work on epitaxial graphene which paralleled the emergence of the graphene research field.[5]

Without appreciating the significance of 2D materials at that time, Loh KP initiated research on MoS_2 (molybdenum disulphide) and h-BN (hexagonal boron nitride) as early as 2004. One rather advanced notion at that time was the synthesis of edge-oriented MoS_2 nanosheets using a single source precursor based on Mo(IV)-tetrakis(diethylaminodithiocarbomato),[10] as opposed to the conventional methods using dual sources of Mo and S. Edge-oriented MoS_2 films exhibit a high density of nanowalls, which had been demonstrated then to exhibit excellent electrochemical charge storage properties.[17] Interestingly, such edge-oriented MoS_2 nanosheet-like films had been found also to exhibit weak magnetism (similar to 1–2 emu/g) and 2.5% magnetoresistance effects with a Curie temperature of 685 K.[18] The magnetisation is related to the presence of edge spins on the prismatic edges of the nanosheets according to spin-polarized calculations.

The growth of h-BN on 6H-SiC (0001) using plasma-excited borazine was studied in 2005 with PhD student Chen Wei. On 6H-SiC (0001), the growth of a pin-hole free, compact h-BN film was difficult due to poor wetting properties between h-BN and 6H-SiC.[19] The strained BN layer would release its elastic energy due to a morphological instability at the interface. This strain relief mechanism gave rise to a buckling of the film into longitudinal islands and round trenches between 500–700°C. The work was selected as the Editor's choice in *Physica Status Solidi*. The expertise developed in the growth of h-BN film at that time was useful to later research in 2D materials, especially in view of the role of h-BN as a dielectric and passivation layer in 2D research.

Loh KP's entry into graphene research started in 2007 when he was encouraged by A Wee to organise a National Research Foundation (NRF) Competitive Research Programme (CRP) project on graphene. His team won the inaugural CRP award in 2008 centred on the theme of Graphene and Related Materials. At the end of this project, more than 10 *Nature* series papers were published as a result of the efforts, ranging from the use of graphene as a broadband polariser to the face-to-face transfer growth of graphene.[20–29] Not forgetting his first love, Loh KP relished the opportunity to marry diamond with graphene, which he achieved by attaching a layer of graphene on diamond and fusing the interface by desorbing the surface hydrogen at a high temperature.[29] To his delight, it was discovered that residual water from the transfer process trapped in the bubbles formed by graphene on diamond became superheated at 600° C. This table-top hydrothermal anvil had allowed the dynamics of supercritical water entrapped between a graphene membrane and diamond to be studied using optical spectroscopy methods. Amazingly, superheated water could be rendered highly corrosive and etches diamond. This work captured the imagination of the local and international media and was reported extensively in the *Sunday Times*, *Today* and *Lian He Zao Bao*. In 2010, Loh KP met AH Castro Neto and got involved in the founding of the Graphene Research Centre.

From Graphene to 2D Materials/GRC to CA2DM (2010-Present) [AH Castro Neto]

In November 2008, Prof Barbaros Oezyilmaz invited me (AH Castro Neto) to come to Singapore to give a talk at the Asia Nano conference that he was helping to organize. At that time graphene research was at its peak with many papers being written every day and, at least in the USA where I was a Professor, we had conferences with thousands of participants. It was obvious to some of us that graphene was going to lead to a Nobel Prize because of the overwhelming response from the whole scientific community. I remember that in 2005, soon after the publication of the famous Manchester Group Science paper on the "scotch tape" exfoliation of graphite, I met Prof Andre Geim personally at the March Meeting of the American Physical Society in Los Angeles. I told my two main collaborators at the time, Prof. Francisco Guinea from Spain and Prof. Nuno Peres from Portugal, that graphene would become the subject of a Nobel prize.

Singapore was already on the world map for graphene with the efforts of Prof Barbaros Oezyilmaz (NUS), Prof Loh Kian Ping (NUS), Prof Andrew Wee (NUS), Prof Yu Ting (NTU), Prof Shen Ze Xiang (NTU), whom I knew from papers. At the time, Prof Andrew Wee suggested we should apply to the National Research Foundation (NRF), under my leadership, for the creation of a S$ 150 million Research Centre of Excellence (RCE) on carbon based materials (fullerenes, carbon nanotubes, graphene, diamond, etc). My wife and I were really positively impressed with Singapore and decided we should try. I remember having great meetings with NUS President Prof Tan Chorh Chuan and NUS Provost Prof Tan Eng Chye who were enthusiastic supporters of the idea.

During the first semester of 2009 we organized several events in Singapore in order to bring together the group that would seed this new RCE. This group included Kostya Novoselov who, at the time, was still a Royal Society Research Fellow in Manchester and who I had convinced to move to Singapore. In the second semester of 2009 I became a Visiting Professor at NUS and worked on the proposal for the new RCE. Unfortunately that year the RCE funding was terminated. By the end of my visit in December 2009, on my way to become a

Visiting Miller Professor in Berkeley, I said my goodbyes to Singapore and thanked everybody who gave support to the project. During my stay in Berkeley I received a phone call from Prof Tan Eng Chye who proposed the creation of a smaller research centre, with "only" S$ 40 million, focused on graphene. One has to keep in mind that this was before the Nobel Prize and I give a lot of credit to the NUS administration regarding their vision that graphene would play an important role in science and technology. That night I remember calling two key people to discuss Prof Tan Eng Chye's proposal, Prof Andre Geim and Prof Philip Kim (at the time at Columbia University), and both were very supportive of the idea. The support from the key graphene people was fundamental to my decision to move forward and take leave of absence from Boston University in the USA and move to Singapore in August 2010.

With the guidance of Prof Novoselov and Prof Oezyilmaz and the outstanding assistance of Dr Peter Blake (Manchester University) and Mr Ang Han Siong (Graphene Research Centre's facilities manager), together with the invaluable administrative support by Ms Lee Wei Fen, we took the steps towards the creation of the Graphene Research Centre (GRC) micro and nano-fabrication facility which was envisioned to be the first of its kind in the world completely focused on graphene. The experience of creating this facility was daunting given the lack of expertise in Singapore in building high technology facilities of this nature. However, the support of the Faculty of Science (FoS), at the time under the leadership of Dean Andrew Wee was extraordinary. We were so lucky to work with people such as Mr Syam Kumar Prabhakaran and Ms Belinda Beh Hui Min, who not only facilitated our life tremendously but who were also directly involved in all the steps that would lead to a world class facility.

By September 2010 we had made all the arrangements for Prof Novoselov to move to Singapore. However, the rumours of a Nobel Prize for Geim, Novoselov and Kim made me worry a little bit but I tried to bring myself to think that it was too soon for a Nobel Prize, but that would mean Kostya would be awarded it as a Professor at NUS, which would have been a major coup for NUS. During an award ceremony in early October 2010 I recalled telling Prof Tan Eng Chye and Prof Andrew Wee that I was crossing my fingers that they would not get the Nobel Prize that year so that Kostya would join us at NUS. I was convinced that the United Kingdom would not let him leave if he had won the Nobel Prize while still at Manchester.

On October 6th 2010, Prof Vitor Pereira (NUS) and I were in my office at NUS on a teleconference with Kostya discussing a joint project when Kostya phone rang. The announcement had been made and Kostya together with Andre had been awarded the 2010 Nobel Prize in Physics. We were witnesses of the phone call that had brought Kostya the good news. For me it was a bitter sweet experience since

Fig. 8. [Top] Prof Novoselov, Prof Castro Neto, and Prof Geim at the Nobel Ceremony, December 2010, Stockholm, Sweden. [Bottom] Prof Geim and Prof Novoselov receiving the 2010 Nobel Prize in Physics from the King of Sweden.

I was happy for my friends but realised that to bring Kostya to Singapore was now essentially out of the question. My own consolation was that Kostya invited me and my wife for the Nobel Prize award ceremony in Stockholm in December 2010 which was an unforgettable experience (Figure 8).

Although Kostya did not move to NUS, he continued to play a big role in helping us build our facilities. It took almost one year of planning plus one year of actual construction and in June 2012 our nano-fabrication facility was ready. We had the honour with the visit of Deputy Prime Minister Teo Chee Hean who experienced, under the guidance of Prof Novoselov, the thrill of producing a graphene device

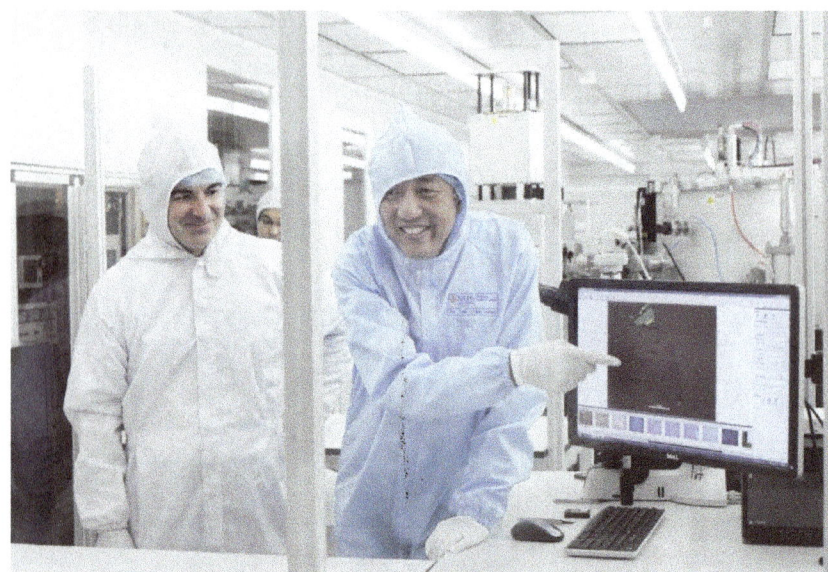

Fig. 9. Prof Novoselov and DPM Teo at the GRC cleanroom after DPM Teo produced his first graphene flake under Prof Novoselov's guidance.

inside our cleanroom (Figure 9). In June 12, 2012, Prof Tan Chorh Chuan inaugurated our world class facility (Figure 10).

Prof Andre Geim has also been a constant presence at GRC since he became Distinguished Visiting Professor at NUS. Since 2012 Prof. Geim has visited GRC one month per year (Figure 11). During his visits Prof Geim acted as a mentor to many of our assistant professors, their research fellows and students. Moreover, collaborations had been born out of these visits and a memorandum of understanding had been signed between Manchester University and NUS.

Since 2005 after the Manchester group published the first paper demonstrating that other 2D crystals besides graphene could be exfoliated by the "scotch tape" method, it was clear that graphene was not alone and that a very large class of layered materials could be exfoliated. Many of these materials, unlike graphene, have many-body electronic states that are not trivial. Transition metal dichalcogenides, for instance, can be charge density waves (CDW) and superconductors, layered cuprate oxides can be high temperature superconductors, layered manganites can be ferromagnets and anti-ferromagnets. Furthermore, graphene is a semi-metal, that is, a metal with a very low density of states, and hence conductive. For that reason, graphene cannot be readily used in digital applications. It had been rather clear that for applications in modern electronics we needed 2D semiconductors and there is a large family of those with different lattice structures, different gap sizes, and different electronic mobilities. Already in 2005 we had known that progress

Fig. 10. Inauguration of GRC Micro and Nano-Fabrication facility in June 2012. In the picture from left to right: FoS Dean Prof Andrew Wee, Prof Castro Neto, Prof Novoselov, Deputy President for Research and Technology Prof Barry Halliwell, NUS President Prof Tan Chorh Chuan and NRF CEO Dr Francis Yeoh.

GRC Christmas party on December 2013: Prof Barbaros Oezyilmaz and Prof Geim.

in this area would be in the production and synthesis of new 2D materials with properties complementary to those of graphene. The expansion of GRC's research scope into new 2D materials was therefore a necessary condition for keeping a worldwide leadership role in materials science and technology. In fact, GRC was the first centre to publish on the electronic properties of phosphorene, a 2D allotrope of phosphorus, which has semiconducting properties, and can be transformed into a metal and even a superconductor under pressure.

Since 2010, researchers at GRC have been very successful in attracting funding and NRF has always been a major funder of GRC activities. In 4 years, GRC researchers were able to attract around S$ 50 million in funding. As a result, hundreds of papers and invention disclosures were written in this area establishing NUS and Singapore as a hub for graphene research. Nevertheless, the costs of running a high tech facility with state of the art characterization, nano-lithography, and device development were very high. Grant funding could not be used to cover operating costs and hence, there was the important issue of the long term sustainability, and even existence, of GRC.

In 2014 NRF awarded NUS a Mid-Size Centre grant of S$ 50 million over 10 years in order to create a natural progression for GRC to reposition itself as the Centre for Advanced 2D Materials, CA2DM. The new funding essentially allowed the continued operation of the high tech facilities that were created in the last 5 years and the retention of the skilled manpower that was needed for such endeavours. Furthermore, the Office for Industry and Innovation, under the leadership of Ms Tricia Chong (NUS Enterprise) was created to facilitate the interaction between academics and the industry.

Today, after 5 years of operation, CA2DM and GRC are centres of reference in the area of materials science and technology. Visitors from all over the world come to Singapore to work in CA2DM's facilities and interact with CA2DM researchers. In a recent *Nature Nanotechnology* editorial,[30] it was written:

> "Nanotechnology in Singapore is taken seriously. In 2010, the National University of Singapore provided S$40 million to set up the Graphene Research Centre, which boasts state-of-the-art fabrication facilities and is home to a faculty featuring several world-renowned experts in both experimental and theoretical techniques. In 2014, the centre expanded to become the Centre for Advanced 2D Materials, thanks to a S$50 million grant from the National Research Foundation (http://go.nature. com/ykV7fZ). This move strengthened the technical capabilities and human resources of the centre and reflected the current trends in research now embracing all types of layered materials beyond graphene, such as transition metal dichalcogenides and black phosphorus."

At CA2DM we do not believe in predicting the future, we believe in inventing it.

References

1. As quoted in *Growth, Dissolution, and Pattern Formation in Geosystems* (1999) by Bjørn Jamtveit and Paul Meakin, p. 291.
2. http://www.physics.nus.edu.sg/~surface/
3. National Science and Technology Board (NSTB), the predecessor of the Agency for Science, Technology and Research (A*STAR)
4. http://ssls.nus.edu.sg/
5. W. Chen, H. Xu, L. Liu, X. Y. Gao, D. C. Qi, G. W. Peng, S. C. Tan, Y. P. Feng, K. P. Loh, A. T. S. Wee, Atomic structure of the 6H-SiC(0001) nanomesh, *Surface Science* **596** (2005) 176–186.
6. K. S. Novoselov, A. K. Geim, S. V. Morozov, D. Jiang, Y. Zhang, S. V. Dubonos, I. V. Grigorieva, A. A. Firsov, Electric Field Effect in Atomically Thin Carbon Films, *Science* **306** (2004) 666–669.
7. Nature News 25 March 2009; *Nature* 458, 390–391 (2009).
8. http://www.nature.com/news/2009/090325/full/458390a.html
9. T. Lin, GY Yu, ATS Wee, ZX Shen and KP Loh, Compositional mapping of the argon-methane-hydrogen system for polycrystalline to nanocrystalline diamond film growth in a hot-filament chemical vapor deposition system, *Applied Physics Letters*, 77, 17, 2692 (2000).
10. Chong, KF, Loh, KP, Vedula, SRK, Lim, CT, Sternschulte, H, Steinmuller, D, Sheu, FS, Zhong, YL, Cell adhesion properties on photochemically functionalized diamond, *Langmuir*, 23, 5615–5621(2007).
11. Zhong, YL; Chong, KF; May, PW; Chen, ZK Loh, KP, Optimizing biosensing properties on undecylenic acid-functionalized diamond, *Langmuir* 23, 5824–5830 (2007).
12. Ang, PK; Loh, KP; Wohland, T; Nesladek, M; Van Hove, E, Supported Lipid Bilayer on Nanocrystalline Diamond: Dual Optical and Field-Effect Sensor for Membrane Disruption, *Advanced Functional Materials* 19, 109–116 (2009).
13. Zhong, YL; Loh, KP; The Chemistry of C-H Bond Activation on Diamond, *Chemistry-An Asian Journal*, 5, 1532–1540 (2010).
14. Zhong, YL; Loh, KP; Midya, A; Chen, ZK, Suzuki coupling of aryl organics on diamond, *Chemistry of Materials*, 20, 3137–3144 (2008).
15. Xie, XN; Lim, R; Li, J; Li, SFY; Loh KP; Atomic hydrogen beam etching of carbon super-structures on 6H-SiC(0001) studied by reflection high-energy electron diffraction, *Diamond and Related Materials*, 10, 1218–1223 (2001).
16. Xie, XN; Wang, HQ; Wee, ATS; K. P. Loh, The evolution of 3 x 3, 6 x 6, root 3 x root 3R30 degrees and 6 root 3 x 6 root 3R30 degrees superstructuves on 6H-SiC (0001) surfaces studied by reflection high energy electron diffraction, *Surface Science*, 478, 57–71 (2001).

17. Soon, JM; Loh, KP, Electrochemical double-layer capacitance of MoS2 nanowall films, Electrochemical and Solid State Letters, 10, A250–A254 (2007).
18. Zhang, J; Soon, JM; Loh, KP; Yin, JH; Ding, J; Sullivian, MB; Wu, P; Magnetic molybdenum disulfide nanosheet films, *Nano Letters*, 7, 2370–2376 (2007).
19. Chen, W; Loh KP; Lin, M; Liu, R; Wee, ATS; Atomic force microscopy study of hexagonal boron nitride film growth on 6H-SiC (0001) — Editor's choice, *Physica Status Solidi A-Applications and Materials Science*, 202, 37–45 (2005).
20. Gao, Libo; Ni, Guang-Xin; Liu, Yanpeng, A. H. Castro-Neto, Kian Ping Loh, Face-to-face Transfer of Wafer-Scale Graphene Films, *Nature*, 505, 190–194 (2014).
21. Manish Chhowalla, Goki Eda, Kian Ping Loh et al., The chemistry of ultra-thin transition metal dichalcogenide nanosheets, *Nature Chemistry* 5, 263–275 (2013).
22. Jian Zheng and Kian Ping Loh, High Yield Exfoliation of Two Dimensional Chalcogenides using Sodium Naphtalenide, *Nature Communications*, 5, 2995 (2014).
23. Jiong Lu, Pei Shan Emmeline Yeo, Chee Kwan Gan, Ping Yu and Loh, KP, Transforming Fullerene Molecules into Graphene Quantum dots, *Nature Nanotechnology*, 6, 247–252, (2011).
24. Q. Bao, H. Zhang, B. Wang, Z. Ni, Candy H.Y.X. Lim, Y. Wang, D. Y. Tang, Loh KP, Graphene as broadband polarizer, *Nature Photonics*, 5, 411–415 (2011).
25. Loh KP, Bao QL, Eda G, Manish Chowalla, Graphene oxide as a chemically tunable platform for optical applications, *Nature Chemistry*, 2, 12, 1015–1024 (2010).
26. Jiong Lu, A. H. Castro Neto and Kian Ping Loh, Transforming Graphene Moire Blisters to Geometric Nanobubbles, *Nature Communications* 8, 823 (2012).
27. Jiong Lu, Kai Zhang, Tze Chien Su, A. H. Castro Neto, Kian Ping Loh, Order-Disorder Transition in a 2-D B-C-N alloy, *Nature Communications* 4, 2681 (2013).
28. Chen Liang Su and Kian Ping Loh et al., Probing the Catalytic Activity of Graphene Oxide and its origin *Nature Communications*, 3, 1298 (2012).
29. Candy Su, Kian Ping Loh et al., A hydrothermal Anvil made of Graphene nanobubbles on diamond *Nature Communications* 4, 1556 (2013).
30. Editorial, *Nature Nanotechnology*, Vol. 10, October 2015, p. 825.

8. Electronic Materials Research in Singapore

Chee Ying Khoo*, Pooi See Lee*, Sze Ter Lim† and Chee Lip Gan*

In the early 1960s, Singapore had the only TV assembly plant in Southeast Asia. By 2012, electronics manufacturing contributed 5.2% to Singapore's Gross Domestic Product (GDP). Today, Singapore's electronics industry plays a crucial role in the global electronics market. It has become the choice location for electronics companies to build and explore new markets. Singapore is currently home to 14 silicon wafer fabrication plants, around 20 semiconductor assembly and test companies, 15 of the world's top 25 fabless semiconductor companies, 11 of the world's top 20 integrated device manufacturers, the world's top 3 hard disc drive manufacturers, and 4 of the world's top 5 Electronic Manufacturing System (EMS) providers.

10 of the 14 Silicon wafer fabrication plants in Singapore have the 200-mm facility and 4 of them (GlobalFoundries, United Microelectronics Corporation, Micron) own 300-mm facility. Chartered Semiconductor (currently known as GlobalFoundries Singapore) completed the first 200-mm fab in 1996 in Singapore and started commercial shipment from the first 300-mm facility in 2005.

In 1991, the Institute of Microelectronics (IME), was one of the research institutes that was formed under the Agency for Science, Technology and Research (A*STAR). IME is committed to contribute to the microelectronics industry in Singapore by undertaking Research and Development (R&D) in microelectronics and also supporting the R&D needs in the industry. IME is also playing a crucial role in developing skilled R&D personnel. Besides IME, the Data Storage Institute (DSI) and Institute of Materials Research and Engineering (IMRE), A*STAR, established in 1996 and 1997 respectively, are also involved in various aspects of electronics materials research in Singapore. Besides the research institutes under A*STAR, faculty at both Nanyang

---

*School of Materials Science & Engineering, Nanyang Technological University.
†Data Storage Institute, A*STAR, Singapore.

Technological University (NTU) and the National University of Singapore (NUS) have been involved in different areas of electronics materials research over the past decades.

In 1998, NTU, NUS and Massachusetts Institute of Technology (MIT) together set up the Singapore-MIT Alliance (SMA) collaboration, which involves innovative engineering education and research. In particular, the Advanced Materials for Micro- and Nano-Systems (AMM&NS) programme under the SMA initiative focused on advance materials, with emphasis on applications in microelectronics and emerging nano-technologies.

Singapore has invested heavily in R&D. The government raised the R&D budget from 2011–2015 by 20% from the previous phase (2006–2010). Such an increase indicates the government's aim to build Singapore into one of the most research-intensive countries in the world. Singapore's current well-developed electronic R&D capabilities range from materials development, component-level design and process, system-level product design, firmware development and also industrial design.

In 2011, NUS and A*STAR jointly launched research programmes in the new growth area of green electronics. It is undertaken by the Department of Electrical and Computer Engineering (ECE) at NUS Faculty of Engineering in collaboration with A*STAR's DSI and IME.

In 2012, IME together with Applied Materials officially opened the Centre of Excellence in Advanced Packaging, with a combined investment of over USD 100m. It has a world-class facility that features a 14,000 square foot Class-10 cleanroom and is equipped with a fully-integrated line of 300-mm manufacturing systems that are able to support the research and development of 3D chip packaging. This initiative is expected to contribute and accelerate the development and adoption of 3D packaging technology globally. In the same year, IMRE together with Cima NanoTech, a US multinational company, signed an agreement to work on new sustainable nanomaterials in order to make cheaper and more efficient electronics and organic solar cells.

In 2014, IME and its 10 industry partners formed 4 joint laboratories, with a total commitment of S$200m. The Advanced Semiconductor Joint Labs were formed to develop and advance semiconductor technologies for future electronic markets. The 4 joint laboratories — in lithography, wafer level packaging (WLP), metrology and assembly — are able to provide an integrated platform for R&D in the semiconductor field. They involve patterning, further development of 3D Integrated Circuits (ICs), quality control and the assembly and high-volume manufacturing of chips.

In this chapter, a summary of some of the key developments that have been achieved locally will be highlighted, focusing on: front-end-of-line (FEOL) materials such as gate dielectrics, contacts and silicides; back-end-of-line (BEOL) materials such as interconnects, inter-level dielectrics and packaging materials; and on to magnetic materials.

1. Front End of Line (FEOL) Materials

A. Dielectric materials

The significant breakthrough in dielectrics in the past five decades is the deployment of alternative dielectric materials to fuel and facilitate the aggressive scaling of complementary metal-oxide semiconductor (CMOS) devices which poses significant leakage current and reliability concerns for the conventional silicon dioxide dielectric materials. HfO_2 is considered the most promising candidate as the first-generation high-k material due to its high dielectric constant (25–30), large energy bandgap (~5.6 eV), high electron offset (1.5 eV) and high breakdown field (13–20 MV/cm). To facilitate the deposition of high quality and homogenous high-k dielectric layers, unique atomic layer deposition (ALD) has been developed with suitable precursors for controlled layer by layer deposition with self-limiting growth of high-k dielectrics. For example, investigation has been carried out on the laminate form of HfO_2 and Al_2O_3 deposited using atomic layer deposition.[1] Scaling to 22 nm and below requires a specialized chamber for interface layer control and post deposition annealing. ALD enables high quality dielectrics with uniform coverage even on graphene, giving low leakage current density and maintaining excellent Hall mobility for graphene.[2] This is a paradigm shift in materials processing, considering the conventional approach of thermal oxidation of silicon for the growth of silicon dioxide in the past decades.

Alternative materials or next generation high-k dielectrics have also been explored using various depositions or coating methods on silicon with the aim to provide alternative dielectric materials selection. For example, lanthanide-based high-k dielectrics (or rare earth oxides) have attracted tremendous interest due to their good thermal stability and considerable high dielectric constant, as shown in Fig. 1.[3]

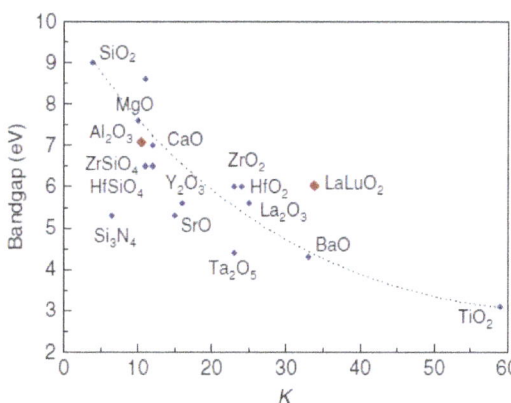

Fig. 1. Static dielectric constant versus bandgap for various high-k dielectrics. Reprinted with permission from J. Robertson.[3] Copyright 2011 Elsevier.

Lee et al. have studied the thermal stability of ultrathin Lu_2O_3 deposited with pulse laser deposition. The effect of N_2 or O_2 annealing on electrical behavior[4] and the impact of high temperature annealing on the formation of interfacial silicate layer have been studied.[5] Other lanthanide based dielectrics with attractive high-k properties include La_2O_3,[6,7] Gd_2O_3,[8,9] $LaAlO_3$[10] and $LaLuO_3$.[11]

The materials' properties such as thermal stability of the selected high-k dielectrics must be taken into account. The introduction of pseudo-binary oxides such as $(HfO_2)_x(SiO_2)_{1-x}$ and $(ZrO_2)_x(SiO_2)_{1-x}$ has been attempted to improve the thermal stability.[12] The capping layer on high-k dielectric film has been shown to increase the crystallization temperature and extending the thermal stability of the dielectric stack, albeit at the trade-off on the permittivity or dielectric constant. A La_2O_3 capping layer on HfO_2 exhibits potential for EOT scaling beyond the 22 nm node that required equivalent oxide thickness (EOT) < 0.6 nm.[13,14] Doping of transition and rare-earth metals, for example, La and Al into HfO_2, has been demonstrated to be promising in achieving thermal stability and threshold voltage control.[15,16]

The development of suitable metal gates is critical with the implementation of high k dielectrics, with the purpose to replace poly-Si as the gate electrode in nano-CMOS. For example, Li et al. have developed HfN/SiO_2 and HfN/HfO_2 gate stack technology[17] using the high thermal stability and oxygen resistance of the HfN metal gate. On the materials' reliability aspect of dielectrics, there was also a rise in attention given to the physical analysis of breakdown in advanced high-k gate dielectrics using atomistic simulations, with a comprehensive model for the breakdown mechanism of HfO_2 high-k gate stacks having been presented by Pey et al.[18] Replacement of the Si channel by alternative channel materials for mobility improvement included Germanium (Ge) for p-type FET and III-V materials like InGaAs for n-type field effect transistor (FET) applications.[19] Integration of HfAlO dielectrics and the TaN metal gate onto the $In_{0.53}Ga_{0.47}As$ n-channel FET has been attempted by Lee et al.[20] High mobility high-k on Ge pMOSFETs with 1 nm EOT has been achieved with the use of the post-gate treatment of Ge interface engineering along with the GeO_2 surface passivation led by Zhu et al.[21] Yeo et al. found that surface passivation is effective for alternative substrates such as gallium arsenide[22] and gallium nitride[23] in the integration of high-k dielectrics.[24]

B. Contact materials

High speed field effect transistors requiring ultrashallow junction, and the recent use of silicon on insulator substrates demanding the formation of thin and conductive contact materials on silicon, have raised the fervent search for suitable contact materials. Silicides, an intermediate compound of metal and silicon, have become essential for the formation of ohmic contacts, local interconnects and even Schottky barriers to reduce the series resistance of electronic devices. During the last decade, refractory metal silicides (group IVB, VB and VIB) and near noble metal silicides (group VIII) have been extensively used in the microelectronics industries. Much effort has been spent on improving the stability of the NiSi phase. Lee et al. reported using NiPt alloy to form a ternary nickel platinum silicide Ni(Pt)Si.[25,26] A Pt interlayer between Ni and Si has been incorporated to achieve enhanced phase and morphological stability of NiSi.[27] The use of Ti-capping on Ni film has been used to assist the NiSi formation by gettering the oxygen atoms from the native or residual oxides at the metal/silicon interface.[28] A similar approach has been done with the incorporation of small amount of Ti into the Ni film. Device integration into multiple gate transistors with a sub-35 nm gate length was shown using novel nickel alloy silicides.[29]

2. Back End of Line (BEOL) and Packaging Materials
A. Metallization materials and reliability

Copper (Cu), with its lower resistivity of 1.7 µΩ-cm as compared to 3.0 µΩ-cm of aluminium (Al), will naturally lower the resistance in the metal line, and reduce the interconnect time delay. At the same time, Cu has the added advantage of having a higher activation energy for lattice and grain boundary diffusion, leading to better resistance to electromigration (EM) failures as compared to Al. EM is a key reliability issue in Cu interconnect metallization. It is the enhanced atomic displacement and the accumulated effect of mass transport under the influence of electrical current. EM in Al involves grain boundary diffusion whereas surface diffusion is dominant in Cu.

Although there are many benefits to be gained from the migration of Al/SiO$_2$ to Cu/low-k dielectrics, there were several process integration challenges that must be addressed. That includes the fast diffusion of Cu in both silicon and oxide, which can lead to increased oxide and junction leakage and also lower oxide breakdown voltage. There is therefore a need for a barrier layer to prevent the diffusion of Cu, and also help Cu to adhere well to the dielectric. The barrier layer has to be continuous to prevent Cu diffusion but thin enough to reduce the resistivity of the metal wire. This becomes more challenging as the dielectric becomes more porous while interconnect dimensions are getting smaller. Different types of barrier layers were investigated and their degradation processes were determined.

Latt et al. studied the impact of the different thickness of the tantalum nitride (TaN) diffusion barrier had, which was prepared using Ionized Metal Plasma (IMP) sputtering.[30] IMP deposition produces a TaN film with good step coverage and having denser microstructure which can effectively block the diffusion and intermixing of Cu through a TaN film. Yang et al. evaluated three different types of barrier layers, namely Tantalum (Ta), TaN and bilayer Ta/TaN deposited on ultra-low-k porous polymer (ULKPP).[31] Their electrical resistance and thermal stability were determined and the bilayer Ta/TaN film seems to be a better candidate as it took advantage of the lower resistance of Ta and higher thermal stability of TaN. Nguyen et al. showed that the SiC:H/Ta barrier could be better than the TaN/Ta barrier structure as it shows a higher electrical breakdown field, indicating a significant improvement in terms of its electrical performance.[32]

Chen et al. investigated the interaction at the interface between the Ta barrier layer and the low-k porous dielectric film.[33] The barrier performance was found to be strongly dependant on the chemical interaction at the interface of the barrier layer and the pore-sealing layer. It was shown that the barrier performance can be improved significantly by nitrogen incorporation during liner growth. Chen et al. also explored the various bilayer barrier layers that were formed by silicon carbide films (SiCO, SiCN, SiC) with Ta.[34] SiC/Ta and SiCN/Ta bilayer barriers show significant performance improvements in terms of breakdown strength and leakage current when compared with conventional metal barriers.

Ti-Si-O-N film was shown by Ee et al. to exhibit high stability against Cu diffusion under bias temperature stressing (BTS) when compared to binary barrier materials.[35] This makes Ti-Si-O-N an attractive candidate to be considered as a diffusion barrier. Ta-Ni amorphous thin films were also proposed to be used as an effective copper diffusion barrier by Yan et al.[36] The incorporation of a few percent of oxygen into the film is effective in retarding Cu diffusion and reaction with Si, which enhances the barrier performance.

For the Cu metallization itself, stress is generated during its multilevel electrodeposition process and Balakumar et al. reported the effect of the stress and grain growth in the seed layer on the properties of the electroplated Cu interconnection layer.[37]

As mentioned, electromigration is a key reliability concern for Cu interconnects and has been investigated by researchers worldwide. Gan et al. examined the reliability of different interconnect trees with the same current density and also varying the distribution of a fixed current density on a particular interconnect tree.[38,39] It was shown that the most highly stressed segments are not necessarily the least reliable ones. This behaviour of Cu is different from that of Al. A multi-segment interconnect tree was used by Chang et al. in which the mortality is governed by interactions among the segments, and not by the stress conditions in individual segments.[40] The presence of an active or passive reservoir changes the EM behavior of an interconnect tree when compared to straight via-to-via lines. A multi-segment interconnect tree provides greater flexibility in testing to reveal the different possible failure mechanisms.

A study by Vairagar et al. showed that the EM behaviour is different for the via-fed upper and lower layer of the dual-damascene test structures.[41] In-situ experiments were carried out and they revealed EM-induced void movement along the Cu/dielectric cap interface, which is the dominant EM path.[42] These effects are specific to Cu dual-damascene structures and these studies led to a better understanding of the real EM mechanisms in dual-damascene Cu interconnects. The EM assessment of Cu interconnects line width down to 100 nm was also carried out which showed that Cu nano-interconnects are promising for mass production.[43]

Fu et al. investigated the reservoir effect of EM reliability in Cu/low-k dual damascene interconnects with different reservoir lengths.[44] The total EM lifetime enhancement in the narrow metal line was found to be higher than the wide metal line. Mario et al. showed a new side reservoir test structure that was able to improve the EM reliability as compared to the conventional end-of-line reservoir structures.[45] The side reservoir had the ability to act as a void trap during EM, and therefore delayed the onset of failure. The study also indicated the presence of pre-existing voids in metal interconnects, which had to be taken into consideration during the process and design of the interconnects to mitigate its effects.

The effect of stress migration (SM) on the EM in the dual damascene Cu interconnect was studied by Heryanto et al.[46] It was found that the EM failure time of the metal line could be strongly affected by the presence of residual stresses due to SM. A failure mechanism model for stress evolution and void formation was proposed to assist in the understanding of the interaction between SM and EM failure mechanisms.

As the majority of interconnects carry pulsed current during operation, Lim et al. performed pulsed current study on the lifetime of Cu/low-k interconnects.[47] It helped in understanding the EM performance under pulsed current conditions to better predict its service EM lifetime. The lifetime was found to increase when compared to that of constant current (D.C.) stressed samples and could be modelled with a modified Black's equation.

B. Low-k dielectrics materials and reliability

Scaling with the technology nodes, the reduction in the cross-sectional dimension of the metal line and dielectric spacing in the back end of line has led to a severe increase in the resistance-capacitance (RC) delay. Low-k dielectric materials were explored to replace the traditional silicon dioxide dielectric in the backend interconnect to improve upon the interconnect (RC) time delay. In addition to improving upon the RC delay, the power consumption,[48,49] the crosstalk between metal lines[48,50] are also reduced with the introduction of low-k materials. Different low-k dielectric candidates with a dielectric constant of less than 2.5 were explored for use in the Cu/low-k system.

Characterization of a photosensitive benzocyclobutene (BCB), a type of spin-on polymer was carried out to understand the curing kinetics and how different curing conditions affect the BCB properties.[51] Yu et al. demonstrated the synthesis and characterization of the silsesquioxane (SSQ) films, with a low dielectric constant of about 2.06.[52] The impact of CVD process conditions on the low-k dielectric (carbon-doped silicon oxide, SiOC) film properties was also investigated by Widodo et al.[53]

Ultra low-k dielectrics with a dielectric constant of less than 2.2 were also studied. Dielectric constants approaching 2.0 were obtained for the nanoporous fluorinated polyimide (FPI) films.[54] The mechanical properties of the low-k films are critical as they affect the devices during chemical-mechanical polishing (CMP) and other integration processes. Lu et al. characterized the low-k dielectric films using nanoindentation with continuous stiffness measurement technique.[55,56] Besides evaluating the mechanical properties of the low-k dielectric films, Wang et al. also carry out adhesion characterization.[57] The thermal property of low-k dielectric (methyl silsesquioxane, MSQ) film was studied by Wang et al. and the compositional and structural changes with temperature were investigated.[58]

Time-dependent dielectric breakdown (TDDB) was also an important failure mechanism in low-k dielectrics. Tan et al. used single line test structures for fast and convenient identification of weak failure points in the Cu/low-k interconnect system, instead of the conventional comb test structures structure.[59,60]

The dielectric degradation mechanism for copper interconnects capped with CoWP was studied and compared with those capped with conventional SiN. The TDDB reliability was improved with the capping of CoWP.[61]

C. 3D interconnects and Cu bonding

Due to the push for increasing the density in ICs, the concept of 3D ICs/interconnects has become increasingly popular. Figure 2 schematically presents the concept of 2D and 3D IC.

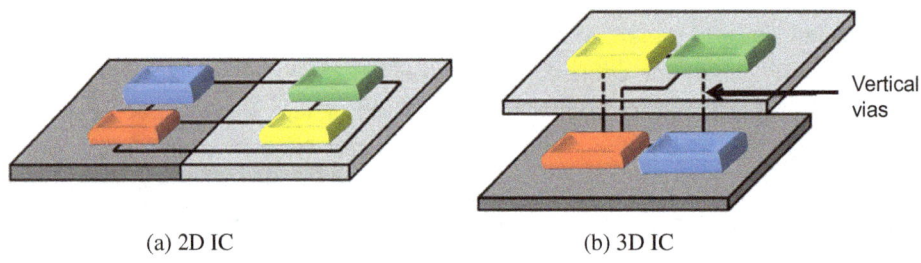

(a) 2D IC (b) 3D IC

Fig. 2. Schematic diagram of 2D and 3D IC.

3D ICs address the scaling challenges faced by 2D dies and connecting them in the third dimension. It is done by stacking silicon wafers and/or dies and interconnecting them vertically using through-silicon via (TSV) in order to achieve improvement in performance with higher bandwidth, reduced power, smaller footprint with shorter interconnect length than the conventional two dimensional processes. They are also able to heterogeneously integrate various functional blocks in a given chip area.

However, there are a number of challenges when implementing the 3D interconnects concept. When Cu-filled TSV is subjected to temperature loadings, there will be a very large local thermal expansion mismatch between the Cu and the Si, as well as between the Cu and the dielectric, and thus creating very large stresses and strains at these interfaces. Nonlinear stresses and strains in the microbumps between the Si chip and Cu-filled TSV interposer were determined for a wide range of via sizes and pitches and at various temperature conditions by Selvanayagam et al.[62] The results obtained are useful for decision making when choosing the underfill materials to minimize the stresses and strains experienced by the microbumps. Finite element modelling was used to evaluate the thermal stresses and strains on the Si due to the proximity of the Cu vias with different geometries.[63]

Annealing silicon wafers with Cu TSVs will cause high stresses in the Cu and protrusions will occur. Heryanto et al. and Che et al. presented a detailed physical characterization of the Cu protrusion effect with its microstructural evolution.[64,65] In order to minimize this Cu protrusion effect, it is proposed to anneal the samples after the electroplating process to avoid further plastic deformation and also help to reduce the diffusive creep.[64] Smaller via volume and less via density designs are desirable to reduce the Cu protrusion and for a given via depth, a smaller diameter via will also decrease the Cu protrusion.[65] A model for Cu protrusion was proposed to understand the failure mechanism, which will be useful to solve TSV-related manufacturing yield and reliability challenges.[64]

Choi et al. reported on the various TSV Cu-filling failure modes and mechanisms.[66] The understanding obtained through this study is extremely useful to overcome the Cu-filling failures and therefore improved the reliability and repeatability of the TSV formation.

TSV architecture suffers from higher power densities which might affect the performance and reliability of 3D ICs if not managed properly. Tan et al. showed the potential benefit of using a Cu bonding medium as a heat flux spreader for overcoming the thermal challenges.[67]

Tan et al. demonstrated the wafer-on-wafer stacking using bumpless Cu-Cu bonding for the simultaneous formation of electrical connection, mechanical support and hermetic frame for an application in 3D ICs.[68,69] A non-thermo-compression bonding method that allows the instantaneous Cu-Cu bonding in room ambient was also proposed, which was able to significantly improve the throughput.[70] Other efforts to lower the required temperature during Cu-Cu bonding such as applying a self assembly monolayer (SAM) onto the Cu surface was also proposed by Ang et al.[71] Lan et al. successfully demonstrated high-density bonded Cu-Cu interconnects using wafer-on-wafer thermocompression.[72,73]

D. Packaging materials

Packaging materials strongly affect the effectiveness of an electronic system. In electronic systems, packaging materials may be used as electrical conductors or insulators, act as thermal paths, and also shield the circuits from surrounding factors, such as moisture, contamination, hostile chemicals, and radiation.

Solders

Lead (Pb)-based solder has been the most commonly used material for bonding purposes in the microelectronics industry but has been prohibited due to environmental and health concerns since 2006. Thereafter, there had been various lead-free solder materials being investigated. High-temperature resistant intermetallic compound (IMC)'s at low bonding temperatures had also been explored by various groups in Singapore.

Attempts were made to lower the bonding temperature. I Made et al. showed that Silver (Ag)-Indium (In) was able to form bonds at 180°C in 10 minutes.[74] Other researchers have shown that Tin (Sn)-In can form good bonding with Cu, and Titanium (Ti) at 180°C.[75–77]

Sn-Ag solders were also investigated for use as an alternative Pb-free solder due to its low bonding temperature.[78,79] Study on the mechanical strength of such thermally aged Sn-Ag/Ni-P[80] and Sn-Ag/Cu solder joints were carried out.[81] Effort was also carried out to create Gold (Au)-Sn solder bumps from electroplating.[82] The thermal-mechanical properties of the Au-Sn alloys were study by Liu et al. to understand the effect of temperature on the hardness and the creep behaviour.[83]

More recently, solder materials for use in ruggedized electronics were explored for use in the deep earth oil exploration, the automotive, and the aerospace industry. These materials need to be able to withstand an extreme operating environment. The mechanical performance of a high temperature Au-Germanium (Ge) made it a suitable candidate as it is able to fulfill the minimum interconnection properties specified by the oil and gas exploration industry.[84,85] A lead-free Ag-Sn soldering technique was also demonstrated to be an effective interconnection method for harsh environment electronic packaging.[86]

Wire bonding

Gold wire bonding is a key technology used in electrical interconnections between the ICs and the external circuitry of microelectronics. The interfacial microstructure evolutions in the Au-Al bond during thermal annealing were carried out.[87,88] Mechanical characterization of the Au wire bonds was also investigated.[89] Dopants such as Calcium (Ca) and Palladium (Pd) were introduced to order to raise the tensile strength and improve the elastic properties of the Au wire.[90,91]

Cu wire bonding technique is more recently explored as an alternative to Au wire bonding due to its lower cost. The interface morphology and metallurgical behavior of the bond formed between the wire and bond pad for different designs and process conditions were studied and Cu bonds were found to be more reliable as compared to Au.[92] Cu wire bonding was shown to be capable of making more reliable electrical interconnects in microelectronic packaging due to its higher electrical conductivity, higher modulus, higher thermal conductivity and lower cost than Au wire.[93–96] The corrosion mechanism in Cu bonding was carried out and guidelines to enhance the bonding resistance were proposed.[97]

There were concerns on corrosion susceptibility and package reliability when using Cu in wire bonding. This motivated the exploration of alternative materials. Pd-coated Cu (PdCu) wire has been proposed as a potential candidate due to its longer shelf life, stitch bond robustness and enhanced bond reliability.[98,99]

Package substrate and encapsulation materials

Substrate materials used for microelectronic packaging need to have a low dielectric constant to minimize signal delay, have a high thermal conductivity so that heat generated can be easily dissipated away, have a closely matched thermal expansion coefficient with Si and have low temperature processibility. Sintered alumina and aluminium nitride (AlN) used to be the common materials for consideration. However, these ceramic substrates are brittle and not easy to be processed. Alternative substrate materials based on composite materials including polymer composites were explored.

Hu et al. proposed alternative substrate materials based on AlN particles and a high temperature resistant engineering thermotropic liquid crystal polymer.[100,101] Dynamic mechanical properties of the polystyrene (PS)-AlN were also examined.[102] Efforts were also carried out to produce AlN microelectronic tapes with a high thermal conductivity and a low dielectric constant.[103] Glass fiber, alumina fiber and carbon fiber reinforced PS composites were found to enhance the thermal properties of the polymer.[104]

IC chips are normally encapsulated with epoxy molding compound (EMC), filled with fused or crystalline silica, which is used to protect the chip and the wire interconnects from the environment and also to provide the required strength. Environmental and viscoelastic effects on the mechanical properties of the epoxy molding compound were investigated by Sung et al.[105] Potential replacements of the conventional fillers with high thermal conductivity ceramic fillers were proposed by Tan et al., which are also environmental friendly by being bromine and antimony free.[106]

Vivek et al. recently evaluated three prospective candidates for high temperature encapsulation, with cyanate ester resin-based composite materials found to be the potential candidate, which could serve as a flip-chip underfill as well as for traditional encapsulation for high-temperature applications.[107,108]

3. Magnetic Materials for Memory

Conventional semiconductor devices have relied on electron charge for both information processing and data storage applications. On the other hand, magnetic recording technology utilizes the electron spin to store the information. Although the latter has a much slower processing speed, it has its merits of non-volatility, better endurance and cost effectiveness.

A. Ferromagnetic semiconductors

Considerable efforts had been made in Singapore to achieve high Curie temperature (T_c) performance as well as to understand the underlying mechanism governing the Curie transition and the physics of the ferromagnetic semiconductor (FMS). Several FMS were fabricated to exhibit ferromagnetic properties by doping semiconductors with transition metals, using various deposition techniques such as pulse laser deposition,[109] molecular beam epitaxy,[110,111] ion implantation[112,113] and sputtering.[114,115] High T_c was achieved in some of these samples including Group IV based FMS, such as Mn ion implanted Si[113] and SiGe[112] ($T_c > 800$ K) and Group II-VI based FMS, $Zn_{1-x}Co_xO$[114] and $Zn_{1-x}Gd_xO$[110] ($T_c \sim 300$ K).

On the other hand, Group IV-VI based FMS, such as $Ge_{1-x}Mn_xTe$, have shown to display high solubility of Mn in GeTe up to 98% grown by MBE, albeit with a low $T_c \sim 95$ K.[110,116] In order to elucidate the physical properties of such a material system various measurements were conducted, including optical, magnetotransport and hydrostatic pressure experiments.[111,117,118] The presence of ferromagnetic signature was verified experimentally and supported by first principle calculations in mostly wide bandgap ZnO and GaN semiconductors, such as Cu doped ZnO[119] and GaN,[120] Carbon doped ZnO,[121] ZnO annealed with hydrogen and Li doped ZnO.[122]

B. Half metals and heusler alloys

Half metals are widely researched in Singapore as well due to their theoretical prediction of a full spin polarization state at the Fermi level (i.e. $P = 1$) which is ideal for an electrode to generate a spin polarized current. Several materials have been studied both theoretically and experimentally, including oxides (Fe_3O_4,[123,124] NiO[125,126]), zinc-blende CrTe[127] and Heusler alloys (Co_2MnAl,[128] Cu_2CrAl,[129] Fe_2CrSi[130–133]). Among which, Heusler alloys, which are well ordered and possess high T_c, are attractive for practical applications. To achieve high spin polarization between ferromagnetic and non-magnetic materials, it is necessary to have a good seed layer and interlayer for epitaxial growth and good band matching at the Fermi surface for a large signal to noise ratio.

The transport properties of the all Heusler GMR devices $Co_2CrSi/Cu_2CrAl/Co_2CrSi$ were also studied using first principle approach based on the density functional theory and the non-equilibrium Green's function method.[134]

C. Magnetic materials for Magnetoresistive Random-Access Memory (MRAM) application

MRAM has attracted immense attention both locally and internationally due to its application as a high density non-volatile memory with commercial products viable within the next five years. The use of perpendicular magnetic anisotropy (PMA) materials for MRAM can be leverage from the knowledge and experiences gained from the research in these materials.

The basis of the PMA based magnetic tunnel junction (MTJ) consists of MgO sandwiched by two thin CoFeB layers. This tri-layer is often grown on a template such as Ta that would reduce roughness and promote a favourable crystal structure that would give rise to a good MTJ performance. However, the diffusion of Ta across the tri-layer can be detrimental to the PMA of the CoFeB and the quality of the MgO tunnel barrier, which was studied by Ying et al.[135] In order to increase the thermal stability, Naik et al.,[136] demonstrated that a thick CoFeB layer up to 2.6 nm with PMA can be achieved. Various materials with a high PMA have also been studied, such as multilayers of Co/Ni,[137] Co/Pd[138] and Co/Pt.[139]

The spin torque efficiency is often used as a metric to account for both the requirements of a low switching current and a high thermal stability which is proportional to the energy barrier.[140] Double MgO structures, i.e. MgO/CoFeB/MgO, have been proposed and have shown to improve the spin torque efficiency.[141,142]

The damping parameter in such systems were measured by Sabino et al.[143] The double MgO structure (MgO/CoFeB/Ta/CoFeB/MgO) has been shown to achieve a high spin torque efficiency to realize practical MRAM devices with low dimensions and high density.[144] Nevertheless, the magnetic materials research in MRAM remain active to discover new magnetic materials with properties tailored to satisfy new device physics, such as the Spin Orbit Torque and the Electric-field based MRAM[145] which brings the promise of faster speed and lower power memory devices.

Indeed, the research on magnetic materials for semiconductor memory application in Singapore saw several achievements and had been very active in the past decades. It continues to be vibrant to meet the insatiable demands for memory devices with higher storage density, multifunctional capabilities, higher endurance and lower power consumption.

4. Outlook and Opportunities

Over the last decades, the silicon-based semiconductor industry strongly adheres to the Moore's Law. However, it is becoming more challenging to follow this paradigm as device scaling has reached both fundamental and economical limits. In order to overcome these limitations, new materials such as compound semiconductors were explored to obtain further integrated circuit performance gain. There is a strong motivation to research on III-V semiconductor materials that are able to integrate with Si. The ability of such integration can therefore lead to the overall improved performance of a system in terms of speed and power consumption.

In 2012, the Low Energy Electronic Systems (LEES) research group under the Singapore-MIT Alliance for Research and Technology (SMART) was started. It aims to identify new integrated circuit technologies that are able to increase the performance and also lower the power consumption in the electronic infrastructure. LEES's proprietary technology on III-V materials growth on Si integration opens up many new opportunities for new circuits and applications such as ultra-efficient circuits for use in remote, mobile and handheld applications. The interconnects between III-V and Si circuits are shorter, which cut down on the interconnect losses. This integration also significantly reduces the chip-footprint, which is able to help in reducing the space required. With such, it is therefore able to accommodate larger batteries and therefore increases the endurance of devices.

LEES perform multidisciplinary research from materials to circuits and systems. LEES is motivated to create a new hybrid III-V on Si CMOS platform to enable new integrated electronic systems. Figure 3 shows the research carried out in LEES, which covers the complete vertical from materials to circuits and systems.

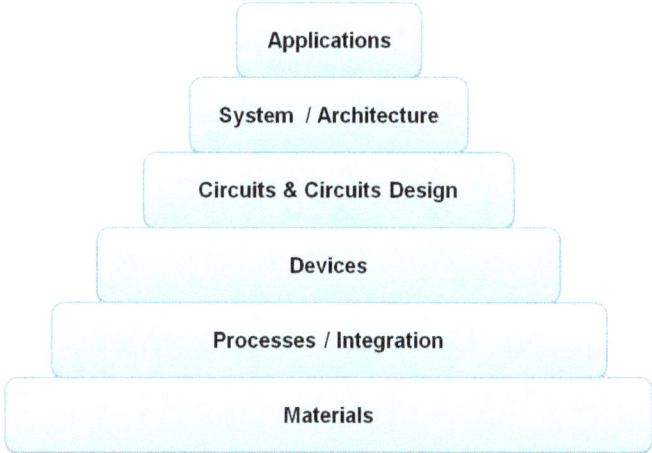

Fig. 3. Multidisciplinary research carried out by LEES.

Future electronic applications in healthcare, automotive industry, mobile devices or human machine interfaces will require thin-film electronic materials. To achieve a full-system integratable with next generation flexible systems, a paradigm shift in design and fabrication is necessary with adjustments from the conventional integrated circuit processes. Recent novel fabrication techniques have allowed thin film transistors to harness the potential of 1-D semiconductors such as Carbon Nanotubes (CNTs) and Nanowires (NWs), as well as 2-D materials such as graphene,[146,147] molybdenum sulfide,[148] and hexagonal boron nitride.[149] Flexible electronics using polymer substrates are capable to create an equivalent of CMOS circuitry, with flexible Radio-frequency identification (RFID) tags being commercialized at an early stage and subsequently wearable electronics for health, lifestyle or sports applications. There is intense interest and focus on the search of emerging electronic materials such as ultrathin black phosphorus (or phosphorene),[150] and 2D van der Waals heterostructures towards spin based electronics.[151] Wet chemical syntheses of graphene and graphene oxides led by Zhang et al., enables high tunability of doping, porosity and functionality. Furthermore, interesting properties of 2-D transition metal dichalcogenide nanosheets have been derived.[152] Materials scientists play a critical role in the progress of electronic materials research and development. While the chemistry and physical characteristics are the overarching properties, electronic materials systems present enormous complexity, also interesting and have diversity in applications, for example, Lee et al. have reported the next generation stretchable transparent conductors for wearable electronics,[153] intrinsically stretchable nanowire photodetectors,[154] foldable nanopaper devices[155] and self-deformable devices.[156]

References

1. Shi-Jin, D., H. Hang, Z. Chunxiang, M.F. Li, S.J. Kim, C. Byung Jin, et al., "Evidence and understanding of ALD HfO2-Al2O3 laminate MIM capacitors outperforming sandwich counterparts," *Electron Device Letters, IEEE*, 2004. vol. 25, pp. 681–683.
2. Zhou, P., S. Yang, Q. Sun, L. Chen, P. Wang, S. Ding, et al., "Direct Deposition of Uniform High-[kgr] Dielectrics on Graphene," *Sci. Rep.*, 2014. vol. 4, 09/29/online.
3. Robertson, J. *New High-K Materials for CMOS Applications. In Comprehensive Semiconductor Science and Technology*, 2011. vol. 4. Amsterdam, The Netherlands: Elsevier.
4. Darmawan, P., C.L. Yuan, and P.S. Lee, "Trap-controlled behavior in ultrathin Lu2O3 high-k gate dielectrics," *Solid State Communications*, 2006. vol. 138, pp. 571–573, 6//.
5. Darmawan, P., P.S. Lee, Y. Setiawan, J. Lai, and P. Yang, "Thermal stability of rare-earth based ultrathin Lu2O3 for high-k dielectrics," 2007.
6. Chin, A. "High quality La2O3 and Al2O3 gate dielectric with equivalent thickness 5–10A," in *VLSI Symposium Technical Digest*, 1999, pp. 135–136.
7. Schamm, S., P. Coulon, S. Miao, S. Volkos, L. Lu, L. Lamagna, et al., "Chemical/Structural Nanocharacterization and Electrical Properties of ALD-Grown La2O3/ Si Interfaces for Advanced Gate Stacks," *Journal of The Electrochemical Society*, 2009. vol. 156, pp. H1–H6.
8. Gottlob, H., M. Schmidt, A. Stefani, M.C. Lemme, H. Kurz, I. Mitrovic, et al., "Scaling potential and MOSFET integration of thermally stable Gd silicate dielectrics," *Microelectronic Engineering*, 2009. vol. 86, pp. 1642–1645.
9. Gottlob, H., T. Echtermeyer, T. Mollenhauer, J. Efavi, M. Schmidt, T. Wahlbrink, et al., "Investigation of high-K gate stacks with epitaxial Gd 2 O 3 and FUSI NiSi metal gates down to CET= 0.86 nm," *Materials science in semiconductor processing*, 2006. vol. 9, pp. 904–908.
10. Lu, X., X. Zhang, R. Huang, H. Lu, Z. Chen, W. Xiang, et al., "Thermal stability of LaAlO3/Si deposited by laser molecular-beam epitaxy," *Applied physics letters*, 2004. vol. 84, pp. 2620–2622.
11. Triyoso, D.H., D.C. Gilmer, J. Jiang, and R. Droopad, "Characteristics of thin lanthanum lutetium oxide high-k dielectrics," *Microelectronic Engineering*, 2008. vol. 85, pp. 1732–1735.
12. Kumar, A., D. Rajdev, and D. Douglass, "Effect of oxide defect structure on the electrical properties of ZrO2," *Journal of the American Ceramic Society*, 1972. vol. 55, pp. 439–445.
13. Ando, T., M. Copel, J. Bruley, M.M. Frank, H. Watanabe, and V. Narayanan, "Physical origins of mobility degradation in extremely scaled SiO2/HfO2 gate stacks with La and Al induced dipoles," *Applied Physics Letters*, 2010. vol. 96, p. 132904.
14. Choi, K., H. Jagannathan, C. Choi, L. Edge, T. Ando, M. Frank, et al., "Extremely scaled gate-first high-k/metal gate stack with EOT of 0.55 nm using novel interfacial

layer scavenging techniques for 22nm technology node and beyond," in *VLSI Technology, 2009 Symposium on*, 2009, pp. 138–139.

15. Park, C., J. Yang, M. Hussain, C. Kang, J. Huang, P. Sivasubramani, et al., "La-doped metal/High-K nMOSFET for Sub-32nm HP and LSTP Application," in *VLSI Technology, Systems, and Applications, 2009. VLSI-TSA'09. International Symposium on*, 2009, pp. 59–60.
16. Lee, P., J. Dai, K.-h. Wong, H.L. Chan, and C.-l. Choy, "Growth and characterization of Hf-aluminate high-k gate dielectric ultrathin films with equivalent oxide thickness less than 10 Å," *Journal of applied physics*, 2003 vol. 93, pp. 3665–3667.
17. Yu, H., H. Lim, J. Chen, M.-F. Li, C. Zhu, C. Tung, et al., "Physical and electrical characteristics of HfN gate electrode for advanced MOS devices," *Electron Device Letters, 2003. IEEE*, vol. 24, pp. 230–232.
18. Ranjan, R., K. Pey, C. Tung, L. Tang, G. Groeseneken, L. Bera, et al., "A comprehensive model for breakdown mechanism in HfO 2 high-k gate stacks," in *Electron Devices Meeting, 2004. IEDM Technical Digest. IEEE International*, 2004, pp. 725–728.
19. Lin, D., G. Brammertz, S. Sioncke, C. Fleischmann, A. Delabie, K. Martens, et al., "Enabling the high-performance InGaAs/Ge CMOS: A common gate stack solution," in *Electron Devices Meeting (IEDM), 2009 IEEE International*, 2009, pp. 1–4.
20. Lin, J., S. Lee, H. Oh, G. Lo, D. Kwong, and D. Chi, "Inversion-mode self-aligned In0. 53Ga0. 47 As N-channel metal-oxide-semiconductor field-effect transistor with HfAlO gate dielectric and TaN metal gate," 2008.
21. Xie, R., T. H. Phung, W. He, Z. Sun, M. Yu, Z. Cheng, et al., "High mobility high-k/ Ge pMOSFETs with 1 nm EOT-New concept on interface engineering and interface characterization," in *Electron Devices Meeting, 2008. IEDM 2008. IEEE International*, 2008, pp. 1–4.
22. Zhu, M., C.-H. Tung, and Y.-C. Yeo, "Aluminum oxynitride interfacial passivation layer for high-permittivity gate dielectric stack on gallium arsenide," *Applied physics letters*, 2006. vol. 89, p. 202903.
23. Liu, X., H.-C. Chin, L.-S. Tan, and Y.-C. Yeo, "In situ Surface Passivation of Gallium Nitride for Metal–Organic Chemical Vapor Deposition of High-Permittivity Gate Dielectric," *Electron Devices, IEEE Transactions on*, 2011. vol. 58, pp. 95–102.
24. Liu, X., H.-C. Chin, E.K.F. Low, W. Liu, L.S. Tan, and Y.-C. Yeo, "In situ Silane Surface Passivation for Gate-First Undoped AlGaN/GaN HEMTs with Minimum Current Collapse and High-Permittivity Dielectric."
25. Lee, P.S., K.L. Pey, D. Mangelinck, J. Ding, D.Z. Chi, J. Dai, et al., "Phase and layer stability of Ni-and Ni (Pt)-silicides on narrow poly-Si lines," *Journal of The Electrochemical Society*, 2002. vol. 149, pp. G331–G335.
26. Lee, P.S., K.L. Pey, D. Mangelinck, J. Ding, T. Osipowicz, and A. See, "Layer inversion of Ni (Pt) Si on mixed phase Si films," *Electrochemical and solid-state letters*, 2002. vol. 5, pp. G15–G17.

27. Liu, J., J. Feng, and J. Zhu, "Film thickness dependence of the NiSi-to-NiSi2 transition temperature in the Ni/Pt/Si (100) system," *Applied physics letters*, 2002. vol. 80, pp. 270–272.
28. Tan, W., K. Pey, S.Y. Chooi, J. Ye, and T. Osipowicz, "Effect of a titanium cap in reducing interfacial oxides in the formation of nickel silicide," *Journal of Applied Physics*, 2002. vol. 91, pp. 2901–2909.
29. Lee, R., L. Tsung-Yang, T. Kian-Ming, A.E.J. Lim, W. Hoong-Shing, L. Poh-Chong, et al., "Novel Nickel-Alloy Silicides for Source/Drain Contact Resistance Reduction in N-Channel Multiple-Gate Transistors with Sub-35nm Gate Length," in *Electron Devices Meeting, 2006. IEDM '06. International*, 2006, pp. 1–4.
30. Latt, K.M., Y.K. Lee, H.L. Seng, and T. Osipowicz, "Diffusion barrier properties of ionized metal plasma deposited tantalum nitride thin films between copper and silicon dioxide," *Journal of Materials Science*, 2001. vol. 36, pp. 5845–51, 12/15.
31. Yang, L.Y., D.H. Zhang, C.Y. Li, and P.D. Foo, "Comparative study of Ta, TaN and Ta/TaN bi-layer barriers for Cu-ultra low-k porous polymer integration," *Thin Solid Films*, 2004. vol. 462–463, pp. 176–81.
32. Nguyen, H.S., Z.H. Gan, C. Zhe, V. Chandrasekar, K. Prasad, S.G. Mhaisalkar, et al., "Reliability studies of barrier layers for Cu/PAE low-k interconnects," *Microelectronics Reliability*, 2006. vol. 46, pp. 1309–14, 08/.
33. Chen, X.T., D. Gui, D.Z. Chi, W.D. Wang, N. Babu, N. Hwang, et al., "Study of Ta-barrier and pore sealing dielectric layer interaction for enhanced barrier performance of Cu/ultralow k(k<2.2) interconnects," *IEEE Electron Device Letters*, 2005. vol. 26, pp. 616–18.
34. Zhe, C., K. Prasad, N. Jiang, L.J. Tang, P.W. Lu, and C.Y. Li, "Silicon carbide based dielectric composites in bilayer sidewall barrier for Cu/porous ultralow-k interconnects," *Journal of Vacuum Science & Technology B (Microelectronics and Nanometer Structures)*, 2005. vol. 23, pp. 1866–72.
35. Ee, Y.C., Z. Chen, T.M. Lu, Z.L. Dong, and S.B. Law, "Low temperature physical-chemical vapor deposition of Ti-Si-N-O barrier films," *Electrochemical and Solid-State Letters*, 2006. vol. 9, pp. 100–3, 03/.
36. Yan, H., Y.Y. Tay, Y. Jiang, N. Yantara, J. Pan, M.H. Liang, et al., "Copper diffusion barrier performance of amorphous Ta-Ni thin films," *Applied Surface Science*, 2012. vol. 258, pp. 3158–62, 01/15.
37. Balakumar, S., R. Kumar, Y. Shimura, K. Namiki, M. Fujimoto, H. Toida, et al., "Effect of stress on the properties of copper lines in Cu interconnects," *Electrochemical and Solid-State Letters*, 2004. vol. 7, pp. 68–71, 04/.
38. Gan, C.L., V.T. Carl, K.L. Pey, W.K. Choi, C.W. Chang, and Q. Guo, "Effect of current distribution on the reliability of multi-terminal Cu dual-damascene interconnect trees," in *International Reliability Physics Symposium, 30 March-4 April 2003*, Piscataway, NJ, USA, 2003. pp. 594–5.
39. Gan, C.L., C.V. Thompson, K.L. Pey, and W.K. Choi, "Experimental characterization and modeling of the reliability of three-terminal dual-damascene Cu interconnect trees," *Journal of Applied Physics*, 2003. vol. 94, pp. 1222–8, 07/15.

40. Chang, C.W., Z.S. Choi, C.V. Thompson, C.L. Gan, K.L. Pey, W.K. Choi, et al., "Electromigration resistance in a short three-contact interconnect tree," *Journal of Applied Physics*, 2006. vol. 99, pp. 94505–1, 05/01.
41. Vairagar, A.V., S.G. Mhaisalkar, and A. Krishnamoorthy, "Electromigration behavior of dual-damascene Cu interconnects-structure, width, and length dependences," *Microelectronics Reliability*, 2004. vol. 44, pp. 747–54, 05/.
42. Vairagar, A.V., S.G. Mhaisalkar, A. Krishnamoorthy, K.N. Tu, A.M. Gusak, M.A. Meyer, et al., "In situ observation of electromigration-induced void migration in dual-damascene Cu interconnect structures," *Applied Physics Letters*, 2004. vol. 85, pp. 2502–2504.
43. Roy, A., R. Kumar, T. Cher Ming, T.K.S. Wong, and C.T. Tung, "Electromigration in damascene copper interconnects of line width down to 100 nm," *Semiconductor Science and Technology*, 2006. vol. 21, pp. 1369–72.
44. Fu, C.M., C.M. Tan, S.H. Wu, and H.B. Yao, "Width dependence of the effectiveness of reservoir length in improving electromigration for Cu/Low-k interconnects," *Microelectronics Reliability*, 2010. vol. 50, pp. 1332–5.
45. Hendro, M., C.L. Gan, Y.K. Lim, J.B. Tan, J. Wei, T. Chookajorn, et al., "Effects of side reservoirs on the electromigration lifetime of copper interconnects," in *2011 18th IEEE International Symposium on the Physical and Failure Analysis of Integrated Circuits (IPFA 2011), 4–7 July 2011*, Piscataway, NJ, USA, 2011, p. 4.
46. Heryanto, A., K.L. Pey, Y.K. Lim, W. Liu, N. Raghavan, J. Wei, et al., "The effect of stress migration on electromigration in dual damascene copper interconnects," *Journal of Applied Physics*, 2011. vol. 109.
47. Lim, M.K., J. Lin, Y.C. Ee, C.M. Ng, J. Wei, and C.L. Gan, "Experimental characterization and modelling of electromigration lifetime under unipolar pulsed current stress," *Microelectronics Reliability*, 2012. vol. 52, pp. 1553–8, 08/.
48. Cregut, C., G. Le Carval, and J. Chilo, "Low-k dielectrics influence on crosstalk: electromagnetic analysis and characterization," in *Proceedings of the IEEE 1998 International Interconnect Technology Conference, 1–3 June 1998*, New York, NY, USA, 1998, pp. 59–61.
49. Baklanov, M., K. Maex, and M. Green, *Dielectric Films for Advanced Microelectronics* Hoboken: John Wiley & Sons, Ltd, 2007.
50. Servel, G., D. Deschacht, F. Saliou, J.L. Mattei, and F. Huret, "Impact of low-k on crosstalk [deep sub-micron technologies]," in *Quality Electronic Design, 2002. Proceedings. International Symposium on*, 2002, pp. 298–303.
51. Chan, K.C., M. Teo, and Z.W. Zhong, "Characterization of low-k benzocyclobutene dielectric thin film," *Microelectronics International*, 2003. vol. 20, pp. 11–22.
52. Yu, S., T.K.S. Wong, X. Hu, and K. Pita, "Synthesis and characterization of porous silsesquioxane dielectric films," *Thin Solid Films*, 2005. vol. 473, pp. 191–195.
53. Widodo, J., W. Lu, S.G. Mhaisaljar, L.C. Hsia, P.Y. Tan, L. Shen, et al., "Characterization of tetra methyl cyclo tetra siloxanes-based low-k dielectric film," *Thin Solid Films*, 2004. vol. 462–463, pp. 213–18.
54. Chen, Y.W., W.C. Wang, W.H. Yu, E.T. Kang, K.G. Neoh, M.H. Vora, et al., "Ultra-low-k materials based on nanoporous fluorinated polyimide with well-defined

pores via the RAFT-moderated graft polymerization process," *Journal of Materials Chemistry,* 2004. vol. 14, pp. 1406–1412.
55. Lu, S., and K. Zeng, "Comparison of mechanical properties of porous and non-porous low-k dielectric films," *Microelectronic Engineering,* 2004. vol. 71, pp. 221–8, 02/.
56. Lu, S., Z. Kaiyang, W. Yihua, B. Narayanan, and R. Kumar, "Determination of the hardness and elastic modulus of low-k thin films and their barrier layer for microelectronic applications," *Microelectronic Engineering,* 2003. vol. 70, pp. 115–24, 10/.
57. Wang, Y.H., M.R. Moitreyee, R. Kumar, S.Y. Wu, J.L. Xie, P. Yew, et al., "The mechanical properties of ultra-low-dielectric-constant films," *Thin Solid Films,* 2004. vol. 462–463, pp. 227–30.
58. Wang, C.Y., Z.X. Shen, and J.Z. Zheng, "Thermal cure study of a low-k methyl silsesquioxane for intermetal dielectric application by FT-IR spectroscopy," *Applied Spectroscopy,* 2000. vol. 54, pp. 209–213.
59. Tan, T.L., C.L. Gan, A.Y. Du, C.K. Cheng, C.M. Ng, and L. Chan, "Test structure design for precise understanding of Cu/low-k dielectric reliability," in *2007 IEEE International Reliability Physics Symposium Proceedings, 15–19 April 2007,* Piscataway, NJ, USA, 2007, p. 2.
60. Tan, T.L., C.L. Gan, A.Y. Du, Y.C. Tan, and C.M. Ng, "Delamination-induced dielectric breakdown in Cu/low-k interconnects," *Journal of Materials Research,* 2008. vol. 23, pp. 1802–1808.
61. Tan, T.L., C.L. Gan, A.Y. Du, C.K. Cheng, and J.P. Gambino, "Dielectric degradation mechanism for copper interconnects capped with CoWP," *Applied Physics Letters,* 2008. vol. 92, pp. 201916–1, 05/19.
62. Selvanayagam, C.S., J.H. Lau, Z. Xiaowu, S. Seah, K. Vaidyanathan, and T.C. Chai, "Nonlinear thermal stress/strain analyses of copper filled TSV (through silicon via) and their flip-chip microbumps," *IEEE Transactions on Advanced Packaging,* 2009. vol. 32, pp. 720–8, 11/.
63. Selvanayagam, C., Z. Xiaowu, R. Rajoo, and D. Pinjala, "Modeling Stress in Silicon With TSVs and Its Effect on Mobility," *IEEE Transactions on Components, Packaging and Manufacturing Technology,* 2011. vol. 1, pp. 1328–35.
64. Heryanto, A., W.N. Putra, A. Trigg, S. Gao, W.S. Kwon, F.X. Che, et al., "Effect of Copper TSV Annealing on Via Protrusion for TSV Wafer Fabrication," *Journal of Electronic Materials,* 2012. vol. 41, pp. 2533–42.
65. Che, F.X., W.N. Putra, A. Heryanto, A. Trigg, Z. Xiaowu, and C.L. Gan, "Study on Cu Protrusion of Through-Silicon Via," *IEEE Transactions on Components, Packaging and Manufacturing Technology,* 2013. vol. 3, pp. 732–9, 05/.
66. Choi, J.W., O.L. Guan, Y. Mao, H.B.M. Yusoff, J. Xie, C.C. Lan, et al., "TSV Cu filling failure modes and mechanisms causing the failures," *IEEE Transactions on Components, Packaging and Manufacturing Technology,* 2014. vol. 4, pp. 581–587.
67. Tan, C.S., "Thermal characteristic of Cu-Cu bonding layer in 3-D integrated circuits stack," *Microelectronic Engineering,* 2010. vol. 87, pp. 682–5, 04/.

68. Tan, C.S., L. Peng, H.Y. Li, D.F. Lim, and S. Gao, "Wafer-on-Wafer Stacking by Bumpless Cu-Cu Bonding and Its Electrical Characteristics," *IEEE Electron Device Letters*, 2011. vol. 32, pp. 943–5, 07/.
69. Tan, C.S., L. Peng, J. Fan, H. Li, and S. Gao, "Three-dimensional Wafer Stacking Using Cu-Cu Bonding for Simultaneous Formation of Electrical, Mechanical, and Hermetic Bonds," *IEEE Transactions on Device and Materials Reliability*, 2012. vol. 12, pp. 194–200, 06/.
70. Tan, C.S., and G.Y. Chong, "High throughput Cu-Cu bonding by non-thermocompression method," in *2013 IEEE 63rd Electronic Components and Technology Conference (ECTC), 28–31 May 2013*, Piscataway, NJ, USA, 2013, pp. 1158–64.
71. Ang, X.F., J. Wei, Z. Chen, and C.C. Wong, "Enabling low temperature copper bonding with an organic monolayer," *Advanced Materials Research*, 2009. vol. 74, pp. 133–6, /.
72. Lan, P., Z. Lin, F. Ji, L. Hong Yu, L. Dau Fatt, and T. Chuan Seng, "Ultrafine pitch (6 m) of recessed and bonded Cu-Cu interconnects by three-dimensional wafer stacking," *IEEE Electron Device Letters*, 2012. vol. 33, pp. 1747–9, 12/.
73. Lan, P., L. Hongyu, L. Dau Fatt, G. Shan, and T. Chuan Seng, "High-density 3-D Interconnect of Cu-Cu Contacts With Enhanced Contact Resistance by Self-assembled Monolayer (SAM) Passivation," *IEEE Transactions on Electron Devices*, 2011. vol. 58, pp. 2500–6, 08/.
74. Made, R., C. Gan, Y. Li, A. Yu, Y. Seung, J. Lau, *et al.*, "Study of low-temperature thermocompression bonding in Ag-In solder for packaging applications," *Journal of Electronic Materials*, 2009. vol. 38, pp. 365–71, 02/.
75. Yan, L.L., C.K. Lee, D.Q. Yu, A.B. Yu, W.K. Choi, J.H. Lau, *et al.*, "A hermetic seal using composite thin-film In/Sn solder as an intermediate layer and its interdiffusion reaction with Cu," 233 Springer Street, New York, 10013–1578, United States, 2009, pp. 200–207.
76. Yu, D., C. Lee, L.L. Yan, M.L. Thew, and J.H. Lau, "Characterization and reliability study of low temperature hermetic wafer level bonding using In/Sn interlayer and Cu/Ni/Au metallization," *Journal of Alloys and Compounds*, 2009. vol. 485, pp. 444–50, 10/19.
77. Yu, D., L. Yan, C. Lee, W.K. Choi, M.L. Thew, C.K. Fool, *et al.*, "Wafer level hermetic bonding using Sn/In and Cu/Ti/Au metallization," in *2008 10th Electronics Packaging Technology Conference (EPTC 2008), 9–12 Dec. 2008*, Piscataway, NJ, USA, 2008, pp. 767–72.
78. Kumar, A., C. Zhong, S.G. Mhaisalkar, C.C. Wong, T. Poi Siong, and V. Kripesh, "Effect of Ni-P thickness on solid-state interfacial reactions between Sn-3.5Ag solder and electroless Ni-P metallization on Cu substrate," *Thin Solid Films*, 2006. vol. 504, pp. 410–15, 05/10.
79. Chen, Z., M. He, and G. Qi, "Morphology and kinetic study of the interfacial reaction between the Sn-3.5Ag solder and electroless Ni-P metallization," in *Lead-Free Solders and Processing Issues Relevant to Micro-Electronic Packaging, 14–18 March 2004*, USA, 2004, pp. 1465–72.

80. He, M., Z. Chen, and G. J. Qi, "Mechanical strength of thermally aged Sn-3.5Ag/Ni-P solder joints," *Metallurgical and Materials Transactions A (Physical Metallurgy and Materials Science)*, 2005. vol. 36A, pp. 65–75, 01/.
81. B.S.S. Chandra Rao, J. Weng, L. Shen, T.K. Lee, and K.Y. Zeng, "Morphology and mechanical properties of intermetallic compounds in SnAgCu solder joints," *Microelectronic Engineering*, 2010. vol. 87, pp. 2416–2422.
82. Wei, J., "A new approach of creating Au-Sn solder bumps from electroplating," *Crystal Research and Technology*, 2006. vol. 41, pp. 150–153.
83. Liu, Y.C., J.W. R. Teo, S.K. Tung, and K.H. Lam, "High-temperature creep and hardness of eutectic 80Au/20Sn solder," *Journal of Alloys and Compounds*, 2008. vol. 448, pp. 340–3, 01/10.
84. Lau, F.L., R.I. Made, W.N. Putra, J.Z. Lim, V.C. Nachiappan, J.L. Aw, et al., "Electrical behavior of Au-Ge eutectic solder under aging for solder bump application in high temperature Electronics," *Microelectronics Reliability*, 2013. vol. 53, pp. 1581–6.
85. Chidambaram, V., B.Y. Ho, and S. Gao, "Reliability of Au-Ge and Au-Si Eutectic Solder Alloys for High-Temperature Electronics," *Journal of Electronic Materials*, 2012. vol. 41, pp. 2107–17, 08/.
86. Sharif, A., G. Chee Lip, and C. Zhong, "Transient liquid phase Ag-based solder technology for high-temperature packaging applications," *Journal of Alloys and Compounds*, 2014. vol. 587, pp. 365–8, 02/2.
87. Xu, C., T. Sritharan, and S.G. Mhaisalkar, "Thin film aluminum-gold interface interactions," *Scripta Materialia*, 2007. vol. 56, pp. 549–52, 03/.
88. Breach, C.D., and F. Wulff, "New observations on intermetallic compound formation in gold ball bonds: general growth patterns and identification of two forms of Au4Al," *Microelectronics Reliability*, 2004. vol. 44, pp. 973–81, 06/.
89. Shah, M., Z. Kaiyang, and A.A.O. Tay, "Mechanical characterization of the heat affected zone of gold wirebonds using nanoindentation," *Transactions of the ASME. Journal of Electronic Packaging*, 2004. vol. 126, pp. 87–93, 03/.
90. Chew, Y.H., C.C. Wong, C.D. Breach, F. Wulff, S.G. Mhaisalkar, C.I. Pang, et al., "Effects of calcium and palladium on mechanical properties and stored energy of hard-drawn gold bonding wire," *Thin Solid Films*, 2004. vol. 462–463, pp. 346–50.
91. Saraswati, T.S., T. Sritharan, C.I. Pang, Y.H. Chew, C.D. Breach, F. Wulff, et al., "The effects of Ca and Pd dopants on gold bonding wire and gold rod," *Thin Solid Films*, 2004. vol. 462–463, pp. 351–6.
92. Murali, S., N. Srikanth, and C.J. Vath, III, "Effect of wire diameter on the thermosonic bond reliability," *Microelectronics Reliability*, 2006. vol. 46, pp. 467–75, 02/.
93. Murali, S., N. Srikanth, Y.M. Wong, and C.J. Vath, III, "Fundamentals of thermo-sonic copper wire bonding in microelectronics packaging," *Journal of Materials Science*, 2007. vol. 42, pp. 615–23, 01/.
94. Zhong, Z.W., "Overview of wire bonding using copper wire or insulated wire," *Microelectronics Reliability*, 2011. vol. 51, pp. 4–12, 01/.
95. Zhong, Z.W., "Wire bonding using copper wire," *Microelectronics International*, 2009. vol. 26, pp. 10–16,/.

96. Breach, C.D., "What is the future of bonding wire? Will copper entirely replace gold?," *Gold Bulletin,* 2010. vol. 43, pp. 150–68.
97. Zeng, Y., K. Bai, and H. Jin, "Thermodynamic study on the corrosion mechanism of copper wire bonding," *Microelectronics Reliability,* 2013. vol. 53, pp. 985–1001.
98. A.B.Y. Lim, A.C.K. Chang, C.X. Lee, O. Yauw, B. Chylak, and Z. Chen, "Palladium-coated and bare copper wire study for ultra-fine pitch wire bonding," *ECS Transactions,* 2013. vol. 52, pp. 717–30, /.
99. A.B.Y. Lim, A.C.K. Chang, O. Yauw, B. Chylak, G. Chee Lip, and C. Zhong, "Ultra-fine pitch palladium-coated copper wire bonding: Effect of bonding parameters," *Microelectronics Reliability,* 2014. vol. 54, pp. 2555–63, 11/.
100. Hu, X., S. Koh Juay, and P. Hing, "Preparation, structure, and properties of aluminium nitride (AlN) reinforced polymer composites: alternative substrate materials for microelectronic packaging," in *Microelectronic Packaging and Laser Processing, 25–26 June 1997,* USA, 1997, pp. 57–65.
101. Hu, X., and P. Hing, "Dielectric, thermal and mechanical properties of microelectronic packaging materials based on AlN filled thermoplastics liquid crystal polymer," in *Dielectric Material Integration for Microelectronics, 3–8 May 1998,* Pennington, NJ, USA, 1998, pp. 288–98.
102. Yu, S., and P. Hing, "Dynamic mechanical properties of polystyrene-aluminum nitride composite," *Journal of Applied Polymer Science,* vol. 78, pp. 1348–1353, 2000.
103. F.Y.C. Boey, A.I.Y. Tok, and W.J. Clegg, "Porous reaction-sintered AlN tapes for high-performance microelectronics application," *Journal of Materials Research,* 2002. vol. 17, pp. 306–14, 02/.
104. Yu, S., and P. Hing, "Thermal and dielectric properties of fiber reinforced polystyrene composites," *Polymer Composites,* 2008. vol. 29, pp. 1199–1202.
105. Yi, S., Z. Yu, R. Neo, and Y. J. Lee, "Hygro-thermo-mechanical behavior of molding compounds at elevated environments," in *2004 Proceedings — 54th Electronic Components and Technology Conference, June 1, 2004 — June 4, 2004,* Las Vegas, NV, United states, 2004, pp. 180–185.
106. Tan, T.H., N. Mogi, and L.P. Yeoh, "Development of environmental friendly (green), thermally enhanced mold compound (TEMC) for advanced packages," in *International Symposium on Electronic Materials and Packaging (EMAP2000), 30 Nov.-2 Dec. 2000,* Piscataway, NJ, USA, 2000, pp. 160–6.
107. Chidambaram, V., J.R.E. Phua, C.L. Gan, and M.W.D. Rhee, "Cyanate Ester-based Encapsulation Material for High-temperature Applications," *Journal of Electronic Materials,* 2013. vol. 42, pp. 2803–12.
108. Chidambaram, V., B.Y. Ho, Y.S. Chan, and M.W.D. Rhee, "High-temperature endurable encapsulation material," in *2012 IEEE 14th Electronics Packaging Technology Conference (EPTC 2012), 5–7 Dec. 2012,* Piscataway, NJ, USA, 2012, pp. 61–6.
109. Lim, S.T., W.D. Song, K.L. Teo, T. Liew, and T.C. Chong, "The Gd Concentration Dependence of the Magnetic Properties of Room Temperature Ferromagnetic Zno:Gd Semiconductor," *International Journal of Modern Physics B,* 2009. vol. 23, pp. 3550–3555.

110. Chen, W.Q., K.L. Teo, S.T. Lim, M.B.A. Jalil, T. Liew, and T.C. Chong, "Magnetic and transport behaviors in Ge[sub 1−x]Mn[sub x]Te with high Mn composition," *Applied Physics Letters*, 2007. vol. 90, p. 142514.

111. Chen, W.Q., S.T. Lim, C.H. Sim, J.F. Bi, K.L. Teo, T. Liew, et al., "Optical, magnetic, and transport behaviors of Ge[sub 1−x]Mn[sub x]Te ferromagnetic semiconductors grown by molecular-beam epitaxy," *Journal of Applied Physics*, 2008. vol. 104, p. 063912.

112. Ko, V., K.L. Teo, T. Liew, T.C. Chong, T. Liu, A.T.S. Wee, et al., "Correlation of structural and magnetic properties of ferromagnetic Mn-implanted Si[sub 1−x]Ge[sub x] films," *Journal of Applied Physics*, 2008. vol. 103, p. 053912.

113. Ko, V., K.L. Teo, T. Liew, T.C. Chong, M. MacKenzie, I. MacLaren, et al., "Origins of ferromagnetism in transition-metal doped Si," *Journal of Applied Physics*, 2008. vol. 104, p. 033912.

114. Tay, M., Y. Wu, G.C. Han, T.C. Chong, Y.K. Zheng, S.J. Wang, et al., "Ferromagnetism in inhomogeneous Zn[1−x]Co[x]O thin films," *Journal of Applied Physics*, 2006. vol. 100, p. 063910.

115. Dietl, T., T. Andrearczyk, A. Lipińska, M. Kiecana, M. Tay, and Y. Wu, "Origin of ferromagnetism inZn1−xCoxOfrom magnetization and spin-dependent magnetoresistance measurements," *Physical Review B*, vol. 76, 2007.

116. Lim, S.T., C.H. Sim, W.Q. Chen, J.F. Bi, K.L. Teo, T. Liew, et al., "Temperature Dependent Magneto-Transport Studies in Ferromagnetic Ge1-Xmnxte with High Mn Composition," *International Journal of Modern Physics B*, 2009. vol. 23, pp. 3591–3595.

117. Lim, S.T., J.F. Bi, K.L. Teo, F.Y. P.T. Liew, and T.C. Chong, "Effect of hydrostatic pressure in degenerate Ge[1−x]Mn[x]Te," *Applied Physics Letters*, 2009. vol. 95, p. 072510.

118. Lim, S.T., J.F. Bi, K.L. Teo, and T. Liew, "Magnetism and magnetotransport studies in Ge0.9Mn0.1Te," *Journal of Applied Physics*, 2011. vol. 109, p. 07C314.

119. Herng, T.S., D.C. Qi, T. Berlijn, J.B. Yi, K.S. Yang, Y. Dai, et al., "Room-Temperature Ferromagnetism of Cu-Doped ZnO Films Probed by Soft X-Ray Magnetic Circular Dichroism," *Physical Review Letters*, 2010. vol. 105.

120. Wu, R.Q., G.W. Peng, L. Liu, Y.P. Feng, Z.G. Huang, and Q.Y. Wu, "Cu-doped GaN: A dilute magnetic semiconductor from first-principles study," *Applied Physics Letters*, 2006. vol. 89, p. 062505.

121. Pan, H., J.B. Yi, L. Shen, R.Q. Wu, J.H. Yang, J.Y. Lin, et al., "Room-Temperature Ferromagnetism in Carbon-Doped ZnO," *Physical Review Letters*, 2007. vol. 99.

122. Yi, J.B., C.C. Lim, G.Z. Xing, H.M. Fan, L.H. Van, S.L. Huang, et al., "Ferromagnetism in Dilute Magnetic Semiconductors through Defect Engineering: Li-Doped ZnO," *Physical Review Letters*, 2010. vol. 104.

123. Jain, S., A.O. Adeyeye, and C.B. Boothroyd, "Effects of buffer layer on the electronic properties of half-metallic Fe3O4," *Journal of Applied Physics*, 2005. vol. 97, p. 10C312.

124. Tripathy, D. and A.O. Adeyeye, "Giant magnetoresistance in half metallic Fe 3O4 based spin valve structures," *Journal of Applied Physics,* 2007. vol. 101, p. 09J505.
125. Wu, R.Q., G.W. Peng, L. Liu, and Y.P. Feng, "Wurtzite NiO: A potential half-metal for wide gap semiconductors," *Applied Physics Letters,* 2006. vol. 89, p. 082504.
126. Wu, R.Q., Y.P. Feng, Y.F. Ouyang, P. Zhou, and C.H. Hu, "Half-metallic NiO in zinc-blende structure from ab initio calculations," *Journal of Applied Physics,* 2008. vol. 104, p. 046103.
127. Sreenivasan, M.G., J.F. Bi, K.L. Teo, and T. Liew, "Systematic investigation of structural and magnetic properties in molecular beam epitaxial growth of metastable zinc-blende CrTe toward half-metallicity," *Journal of Applied Physics,* 2008. vol. 103, pp. 043908–1, 02/15.
128. Qiu, J.J., G.C. Han, W.K. Yeo, P. Luo, Z.B. Guo, and T. Osipowicz, "Structural and magnetoresistive properties of magnetic tunnel junctions with half-metallic Co2MnAl," *Journal of Applied Physics,* 2008. vol. 103, p. 07A903.
129. Ko, V., G. Han, J. Qiu, and Y.P. Feng, "The band structure-matched and highly spin-polarized Co2CrZ/Cu2CrAl Heusler alloys interface," *Applied Physics Letters,* 2009. vol. 95, p. 202502.
130. Y.-P. Wang, G.-C. Han, H. Lu, J. Qiu, Q.-J. Yap, R. Ji, et al., "Tunnel magnetoresistance effect and interface study in magnetic tunnel junctions using epitaxial Fe2CrSi Heusler alloy electrode," *Journal of Applied Physics,* 2013. vol. 114, p. 013910.
131. Y.-P. Wang, G.-C. Han, H. Lu, J. Qiu, Q.-J. Yap, and K.-L. Teo, "Tunnel magnetoresistance effect in magnetic tunnel junctions using Fermi-level-tuned epitaxial Fe2Cr1−xCoxSi Heusler alloy," *Journal of Applied Physics,* 2014. vol. 115, p. 17C709.
132. Y.-P. Wang, G.-C. Han, H. Lu, J. Qiu, Q.-J. Yap, and K.-L. Teo, "Structural and magnetic properties of epitaxial Heusler alloy Fe2Cr0.5Co0.5Si," *Journal of Applied Physics,* 2014. vol. 115, p. 17C301.
133. Y.-P. Wang, J.-J. Qiu, H. Lu, R. Ji, G.-C. Han, and K.-L. Teo, "Perpendicular magnetic anisotropy in Fe2Cr1 −xCoxSi Heusler alloy," *Journal of Physics D: Applied Physics,* 2014. vol. 47, p. 495002.
134. Bai, Z.Q., Y.H. Lu, L. Shen, V. Ko, G.C. Han, and Y.P. Feng, "Transport properties of high-performance all-Heusler Co2CrSi/Cu2CrAl/Co2CrSi giant magnetoresistance device," *Journal of Applied Physics,* 2012. vol. 111, p. 093911.
135. J.-F. Ying, R. Ji, C.C. Wang, S. Ter Lim, H. Xie, and E.F. Gerard, "Observation of TaB x O y Along the Ta Diffusion Path Through a Boron and Oxygen Containing Tri-layer Structure," *Journal of Materials Engineering and Performance,* 2014. vol. 23, pp. 2795–2800.
136. Naik, V.B., H. Meng, and R. Sbiaa, "Thick CoFeB with perpendicular magnetic anisotropy in CoFeB-MgO based magnetic tunnel junction," *AIP Advances,* 2012. vol. 2, p. 042182.
137. M.P.R. Sabino, M. Tran, C. Hin Sim, Y. Ji Feng, and K. Eason, "Seed influence on the ferromagnetic resonance response of Co/Ni multilayers," *Journal of Applied Physics,* 2014. vol. 115, p. 17C512.

138. Tahmasebi, T., S. N. Piramanayagam, R. Sbiaa, and T.C. Chong, "Influence of Spin Polarizer on the Magnetoresistance, Switching Property, and Interlayer Interactions in Co/Pd Single Spin Valves," *IEEE Transactions on Magnetics*, 2012. vol. 48, pp. 3434–3437.
139. Lim, S.T., M. Tran, J.W. Chenchen, J.F. Ying, and G. Han, "Effect of different seed layers with varying Co and Pt thicknesses on the magnetic properties of Co/Pt multilayers," *Journal of Applied Physics*, 2015. vol. 117, p. 17A731.
140. Sun, J.Z., S.L. Brown, W. Chen, E.A. Delenia, M.C. Gaidis, J. Harms, et al., "Spin-torque switching efficiency in CoFeB-MgO based tunnel junctions," *Physical Review B*, 2013. vol. 88.
141. Jan, G., Y.-J. Wang, T. Moriyama, Y.-J. Lee, M. Lin, T. Zhong, et al., "High Spin Torque Efficiency of Magnetic Tunnel Junctions with MgO/CoFeB/MgO Free Layer," *Applied Physics Express*, 2012. vol. 5, p. 093008.
142. Sato, H., M. Yamanouchi, S. Ikeda, S. Fukami, F. Matsukura, and H. Ohno, "Perpendicular-anisotropy CoFeB-MgO magnetic tunnel junctions with a MgO/CoFeB/Ta/CoFeB/MgO recording structure," *Applied Physics Letters*, 2012. vol. 101, p. 022414.
143. M.P.R. Sabino, S. Ter Lim, and M. Tran, "Influence of Ta insertions on the magnetic properties of MgO/CoFeB/MgO films probed by ferromagnetic resonance," *Applied Physics Express*, 2014. vol. 7, p. 093002.
144. M.P.R. Sabino, S.T. Lim, and M. Tran, "Composition and Annealing Temperature Dependence of Magnetic Properties in MgO/CoFeB/Ta/CoFeB/MgO Films," *IEEE Transactions on Magnetics*, 2014. vol. 50, pp. 1–4.
145. Guchang, H., H. Jiancheng, S.C. Hin, M. Tran, and L.S. Ter, "Switching methods in magnetic random access memory for low power applications," *Journal of Physics D: Applied Physics*, 2015. vol. 48, p. 225001.
146. Y.-M. Lin, K.A. Jenkins, A. Valdes-Garcia, J.P. Small, D.B. Farmer, and P. Avouris, "Operation of graphene transistors at gigahertz frequencies," *Nano letters*, 2008. vol. 9, pp. 422–426.
147. Geim, A.K., and K.S. Novoselov, "The rise of graphene," *Nature materials*, 2007. vol. 6, pp. 183–191.
148. Radisavljevic, B., A. Radenovic, J. Brivio, V. Giacometti, and A. Kis, "Single-layer MoS2 transistors," *Nature nanotechnology*, 2011. vol. 6, pp. 147–150.
149. Song, L., L. Ci, H. Lu, P.B. Sorokin, C. Jin, J. Ni, et al., "Large scale growth and characterization of atomic hexagonal boron nitride layers," *Nano letters*, 2010. vol. 10, pp. 3209–3215.
150. Koenig, S.P., R.A. Doganov, H. Schmidt, A.C. Neto, and B. Oezyilmaz, "Electric field effect in ultrathin black phosphorus," *Applied Physics Letters*, 2014. vol. 104, p. 103106.
151. Balakrishnan, J., G.K. W. Koon, M. Jaiswal, A.C. Neto, and B. Özyilmaz, "Colossal enhancement of spin-orbit coupling in weakly hydrogenated graphene," *Nature Physics*, 2013. vol. 9, pp. 284–287.

152. Chhowalla, M., Z. Liu, and H. Zhang, "Two-dimensional transition metal dichalcogenide (TMD) nanosheets," *Chemical Society Reviews*, 2015. vol. 44, pp. 2584–2586.
153. Yan, C., W. Kang, J. Wang, M. Cui, X. Wang, C.Y. Foo, et al., "Stretchable and wearable electrochromic devices," *ACS nano*, 2013. vol. 8, pp. 316–322.
154. Yan, C., J. Wang, X. Wang, W. Kang, M. Cui, C.Y. Foo, et al., "An intrinsically stretchable nanowire photodetector with a fully embedded structure," *Advanced Materials*, 2014. vol. 26, pp. 943–950.
155. Kang, W., C. Yan, C.Y. Foo, and P.S. Lee, "Foldable Electrochromics Enabled by Nanopaper Transfer Method," *Advanced Functional Materials*, 2015.
156. Wang, J., C. Yan, K.J. Chee, and P.S. Lee, "Highly Stretchable and Self-Deformable Alternating Current Electroluminescent Devices," *Advanced Materials*, 2015. vol. 27, pp. 2876–2882.

"Singaporean" Materials Science: What does the Future Hold?

Subbu Venkatraman*

It has been clear from the earlier chapters that today, materials science research and education are thriving in Singapore. As testament to this, the 2016 Quacerelli-Symonds (QS) rankings for Materials Science for NTU and NUS showed them both in the top 10 world-wide (6th and 8th, respectively) which is a remarkable achievement for a small country such as Singapore. Astonishing progress has been made in the last 50 years, and will continue to be made in the next 50, at an even more accelerated pace.

Where do we believe the breakthroughs will be made? First, in the area of sustainable energy: as is evident from Chapter 5, Singaporean researchers are poised to translate novel materials such as Perovskites into solar energy panels. This will be a notable achievement if all the practical issues such as stability, can be overcome, and given the translation thrust of the funding bodies, this is more likely to happen first in Singapore than elsewhere. This effort has been recognized by the conferment of the Singapore National Academy of Sciences' Young Scientist award to Dr Nripan Mathews of NTU for his work on Perovskite-based solar cell materials.

Related to energy conversion is also the issue of better storage systems. Exciting developments have already been reported for supercapacitor materials, and will continue to gain momentum, as will the development of new battery materials, including materials for novel fuel cells and Li- and Na-ion batteries.

Second, in water remediation. Companies such as Hyflux have already put Singapore on the map for water purification systems, and researchers at NEWRI and elsewhere are poised to make even more significant breakthroughs in the detection of contaminants and their efficient removal from waste water. The materials component of water remediation systems will evolve into some of the best-designed

---

*Professor and Chair, School of Materials Science and Engineering, Nanyang Technological University.

worldwide and we fully expect the coveted Lee Kuan Yew Water Prize to be awarded to a Singaporean entity or individual in the next few years.

What of the exciting new world of nanomaterials for medicine? We should expect new nanomaterials such as graphene and nanotubes/nanowires/nanoparticles made out of inorganic and organic materials to start being applied to both diagnose and treat various diseases. Already, much progress has been made in improving treatment options for glaucoma by Singaporean researchers in Nanomedicine: this involves the modification of self-assembling nanoliposomes to slowly release an anti-glaucoma drug in the eye so that one injection can last for several months. This Singaporean achievement was recognized by a President's Technology award in 2014 to a team comprising Professors Freddy Boey and Subbu Venkatraman of NTU and Dr Tina Wong from the Singapore National Eye Centre. An institute, called the NTU-Northwestern Institute for Nanomedicine has been set up based on the use of nanotechnology to diagnose and treat various diseases, including cardiovascular, ocular and skin diseases. The next frontier that would be crossed is likely to be cancer, and in particular, skin cancer. A promising new area involves the use of mesoporous silicone templates for developing photodynamic therapeutics, where localized tumor treatment with minimal side-effects is the goal. A*STAR researchers are also well advanced in the development of functionalized self-assembling nanoparticles for various applications.

Earlier and more robust detection of infectious disease may also be helped by newer nanomaterials, using elegant methods of detection such as plasmonic detection. Therapeutic options for infections and for wound healing will also be advanced by newer materials, and clever use of older materials; one such example is the modification of self-assembling nanoliposomes to slowly release an anti-galucoma drug in the eye so that one injection can last for several months. This Singaporean achievement was recognized by a President's Technology award in 2014 to a team comprising Professors Freddy Boey and Subbu Venkatraman of NTU and Dr Tina Wong from the SIngpaore National Eye Centre. As far as more novel biomaterials are concerned, already hydrogels based on keratin have been developed at NTU, while research is going on apace into tissue-engineered options for skin replacement.

While it might be argued that materials development for tissue-engineered products is already mature, and that further developments will arise more from selection of the right cell types, we believe that soft tissue engineering could still benefit from newer biodegradable scaffold materials. Mimicking the extra-cellular matrix is the best way to go forward, and this will rely substantially on biomimetics to design the optimum scaffold. In fact, considerable synergies are expected between the disciplines of tissue engineering and biomimetic materials. In this context, the exciting new developments in electrospinning at NUS, as well as

3-D printing efforts at NTU, should come together to fabricate elegant scaffolds with outstanding functionalities.

In the field of medical devices, particularly therapeutic devices, fully-degradable materials will continue to draw attention from researchers. As mentioned in Chapter 6, we have barely scratched the surface in terms of medical implants such as stents and occluders. Approvals of such devices will open the floodgates for other applications, particularly in the eye and under the skin. It is obvious that advancements in these spheres will come as a result of newer materials with enhanced functionality or functionalization of well-known materials to improve patient acceptance overall.

Nanocomposites, which combines the disciplines of polymer science, nanomaterials and interfacial chemistry, has a very promising future in Singapore, although the fruits of the research will be applicable on a global scale. Already, companies such as Rolls Royce and Boeing are sponsoring research into aerospace nanocomposites, and their extension into healthcare is only a matter of time. In particular, NTU and the National Dental Centre have teamed up to explore better materials for dental healthcare, and that tie-up is likely to generate a number of novel hard-tissue mimicking biomaterials.

The future of nanocomposites will also feature hybrid and functional nanocomposites. The hybrid nanocomposites, made of inorganic and organic components, holds promise for a number of applications ranging from drug delivery to bio-imaging and solar cell applications. The functional nanocomposite work has applications in conductive high-strength sheets and devices aimed at the flexible electronics market.

Singapore's unique contributions to materials used in water purification was recognized by a President's award in 2015, to Prof Neal Chung of NUS. As detailed in Chapter 3, the Lux report already acknowledges Singapore's high standing in the general area of water remediation research. Future of the membrane work will shift to process improvements as well as into membranes that can better separate organic contaminants from our water supply. Closely tied to the improvements in membrane technology would be the development and scale-up (using green methods) of photocatalytic nanomaterials with improved efficiencies for de-contaminating water. The prototypes will move from the laboratory to the real world in the coming years.

Similar global impact is expected from our work on solar and alternative energy conversion technologies. The two internationally-recognized institutes in Singapore, NTU's ERI@N and NUS' SERIS, will lead the drive toward improved efficiencies of existing solar cell materials and the development of new ones. One particularly exciting new material in this context are the perovskites. Generous funding in this area will surely lead to huge breakthroughs in developing a lead-free and highly efficient perovskite material, that could be used in panels blanketing our skyscrapers for maximum energy extraction.

Similarly, in the area of energy storage, new and exciting nanomaterials are expected to vastly improve efficiencies in batteries, particularly in Li-ion batteries, where Singapore has already made an impact. In the area of supercapacitors, work is aimed at developing Li-ion hybrid electrochemical supercapacitors that will become the power source of choice in electrically-powered vehicles, for example.

In the area of 2-dimensional nanomaterials, Singapore is already recognized internationally as the leading country in the world. The internationally renowned team at NUS at the Centre for Advanced 2D materials (CA2DM) have already won substantial funding and demonstrated the capability to not only expand the range of graphene applications but also to develop newer 2D materials, particularly semi-conducting ones. Such efforts will surely be augmented by NTU's efforts as well into alternative novel 2-D materials.

Electronic materials research is moving more and more towards developing flexible materials and devices. Here the development of not only new materials, but also finding facile routes of synthesis are equally important. Advances in the next generation of stretchable transparent devices are only a matter of when, not if. Wet synthesis routes enable efficient doping as well as control of porosity in the new family of 2-dimensional sheets. The synergy between device development, traditionally a Singapore strength, and new materials synthesis, is an exciting portent for growth in the general area of electronic nanomaterials development.

Index

A
Abbott, 17
AB SCIEX, 17
ACM Biolabs, 17
Activated carbon, 86
Adsorbent materials, 75
Advanced oxidation processes, 75
Advanced oxidation technology, 85
Agency for Science, Technology and Research, 4
Amaranth Medical, 17
Amplified spontaneous emission, 131
Anti-clotting drugs, 167
Anti-fouling, 39, 63
Antimicrobial function, 23
Aquaporins, 61
Atrial septal defects, 169

B
Baxter International, 17
Becton Dickinson, 17
Ben Gurion University, 13
Best Paper Award, 23
biochemical oxygen demand, 49
Biomaterials, 157
biomedical devices, 1
Biomimetic materials, 172
BIOTRONIK, 17
Bisphenol A, 86, 92
Bosch, 143
Boston Scientific, 169
B. subtilis, 83

C
Campus for Research Excellence and Technological Enterprise, 13
Carbon nanotubes, 135
Catalyst materials, 78
Center for Biomimetic Sensor Science, 11
Centre for Advanced 2D Materials, 194
Centre of Excellence in Advanced Packaging, 198
Chemical oxygen demand, 49
Cima NanoTech, 198
Composites, 21
Cranial repair, 170

D
Data Storage Institute, 197
Delamination, 65, 66
Diffuse reflectance infrared fourier transform spectroscopy, 77
Division of Materials Engineering, 6
Dye-sensitized solar cell, 30

E
EcoCampus, 10
E. coli, 83
Economic Development Board, 2, 147
Electric double-layer capacitors, 134
Electronic materials, 197
Energy Research Institute at NTU, 9, 115, 147
Energy storage, 113
Environment and water industry development council, 9

F
Fenton process, 85
Food and Drug Administration, 31
Forward osmosis, 39

G
Gildemeister, 143
GlaxoSmithKline, 17
GlobalFoundries, 197
Graphene, 180
Graphene islands, 180
Graphene research centre, 190
Griddler, 117, 119
Guided bone regeneration, 27

H
Hexagonally packed hollow hoops, 30
Hollow fiber, 56
Hoya Surgical Optics, 17
Hybrid materials, 21
Hydrophilicity, 45
Hydrophobicity, 53
Hyperbranched glycol, 65

I
Ibuprofen, 81
Institute of Materials Research and Engineering, 21, 197
Integrated Circuits, 198
International Finance Corporation, 49

J
Janus nanoparticles, 29
Johnson Matthey, 10
Jurong Industrial Estate, 2
Jurong Town Corporation, 10

L
Layered double hydroxides, 87
Life Technologies, 17
Lithium-ion batteries, 140
Low Energy Electronic Systems, 210

M
Magnetic tunnel junction, 209
Magnetoresistive Random-Access Memory, 209
2D materials, 179
Mechanical characterization, 207
Media development authority, 9
Medical devices, 160
Medlinx Acacia, 17
Medtronic, 17
Membrane bioreactor, 41
Membranes, 37, 77
MEMSTILL, 53
Merck, 17
Methylene blue, 81
Methyl orange, 81
Microfiltration, 38
Ministry of Education, 9
Miscibility, 55
Molecular weight cut-off, 42
Multi-walled carbon nanotubes, 23

N
Nanocomposites, 9, 21
Nanofibers, 84
Nanohybrids, 22
Nanomaterials, 22, 114
Nanorods, 84
Nanyang Environment and water research institute, 9

Nanyang Technological Institute, 4
National Research Foundation, 9, 147
National Science Awards, 180
Novartis, 17

O
organic light-emitting diodes, 9
Organic solar cells, 9
Organoclay, 24
Osteopore, 17
Osteopore International, 170

P
Patent ductus arteriosus, 169
Patent foramen ovale, 169
Peregrine Ophthalmic, 17
Perovskite solar cells, 132
Pfizer, 17
Photocatalytic applications, 76
photocatalytic treatment, 84
Plasmonic effect, 79
Porosity, 55
Porosity gradient, 85
Pressure retard osmosis, 57
PV materials and devices, 114

R
Reactive oxygen species, 79
Reflection high energy electron diffraction, 187
Renewable Energy Corporation, 114
Renewable Energy Integration Demonstrator Singapore, 10
Reverse osmosis, 38
Rhodamine B, 81
Roche, 17

S
Safranin O, 50
Scanning tunneling microscope, 180
Schering-Plough, 17
Schott AG, 120
Secondary Ion Mass Spectrometry, 180
Second Industrial Revolution, 5
SGL Carbon, 143
Siemens Medical Instruments, 17
Silica nanocapsules, 31
SimpliClear, 160, 171
Singapore-MIT Alliance for Research and Technology, 14, 210
Singapore Synchrotron Light Source, 180
Slurry-compounding, 25
Solar energy, 113
Solar Energy Research Institute of Singapore, 115, 147
Spinneret, 55
Staebler-Wronski effect, 119
Stent, 167
Sud-Chemie, 143
Supercritical water, 189

T
Takeda, 17
Temasek Labs, 11
Thin film composite, 46
Time-dependent dielectric breakdown, 204
Tissue engineering, 157
Tissue replacement, 170

U
Ultrafiltration, 38
Ultrahigh vacuum, 180

V
Vapor grown carbon fibers, 31
Vestas, 10

W
Water Remediation, 37, 75
Wyeth, 17

www.ingramcontent.com/pod-product-compliance
Lightning Source LLC
Chambersburg PA
CBHW080410230426
43662CB00016B/2360